The Nature of Physical Geography

2

The Nature of Physical Geography

K.J. Gregory
Professor of Geography, University of Southampton

Edward Arnold

First published in Great Britain 1985 by
Edward Arnold (Publishers) Ltd, 41 Bedford Square, London WC1B 3DQ

Edward Arnold (Australia) Pty Ltd, 80 Waverley Road, Caulfield
East, Victoria 3145, Australia

Edward Arnold, 300 North Charles Street, Baltimore, Maryland 21201,
USA

British Library Cataloguing in Publication Data

Gregory, K.J.
The nature of physical geography.
1. Physical geography
I. Title
910'.02 GB54.5

ISBN 0-7131-6431-X

Text set in 10/11 pt Plantin Compugraphic
by Colset Pte Ltd., Singapore
Printed and bound by Richard Clay (The Chaucer Press) Limited,
Bungay, Suffolk

Contents

Acknowledgements

The publishers would like to thank the following for permission to include figures which are copyright and are used by permission:

American Journal of Science (fig. 8.3) and American Journal of Science and International Association of Scientific Hydrology (fig. 6.1); Edward Arnold (Publishers) Ltd (fig. 3.1); Oxford University Press, Oxford and New York (figs. 4.1, 5.2 and 8.2); Prentice Hall Inc. (figs. 7.1 and 8.1) and John Wiley & Sons Ltd (fig. 5.1). Full citations may be found by consulting the captions and Bibliography.

Preface

This book endeavours to survey how physical geography has developed in recent years but this is confounded by writing from the viewpoint of personal experience, by selection of material from the enormous range available, by the difficulty of ignoring present knowledge when evaluating past developments, by appreciating the context in relation to other sciences, and because there may not be one single explanation of the real physical world. Physical geography is now poised for great development in view of the focus of recent works and research and the advent of new technology including the microcomputer, electronic instrumentation and enhanced remote sensing. Hence it may be even more difficult to envisage writing a book of this kind in future years!

During the writing I have been most grateful to the University of Southampton, who provided a term's periodic leave during which many of the chapters were finalized. A number of experts have been kind enough to read and comment upon particular chapters and I am particularly grateful to Professor J.H. Bird, Professor D.Q. Bowen, Dr M.J. Clark, Dr A.M. Gurnell and Dr D.E. Walling for their endurance in doing this and for the helpful comments which they provided. Grateful thanks are due also to Mr W. Mansfield who proved that it is possible to read the entire book, and to Mr A.S. Burn and his staff in the Cartographic Unit, University of Southampton for the construction of the figures, to Mrs Tina Birring who had grappled with my writing and produced much of the typescript, to Mr P. Burkhard who assisted with the index, to Mrs J. Ghandi who typed it and to Mr P. Boagey, Librarian and Map Curator who has helped with many impossible requests for references. My family has tolerated this exercise as it has lived through the earlier ones and I continue to appreciate their support, their restraint in not asking about completion too often and the brilliant copy corrections suggested by my wife. I would also like to acknowledge the way in which during my undergraduate days Professors E.H. Brown and T.J. Chandler opened my eyes to the breadth and excitement of physical geography and then subsequently Eric Brown guided my postgraduate research. I hope that I have been able to continue the enthusiasm that they conveyed to me and that anyone who reads this text will appreciate something of the excitement that physical geography can generate. This book could, and perhaps should, have taken much longer to write; I hope that it will be read in the light of the knowledge that I already appreciate many of the book's limitations but will be only too pleased to receive comments about those limitations and about others which I have not anticipated.

Ken Gregory
August 1984

Introduction

1

Prologue and presumption

It should be admitted from the beginning that it is presumptuous for one physical geographer to attempt to survey the discipline with equal insight into its major sub-divisions of geomorphology, climatology and biogeography. However, this present book was originally conceived after a suggestion by my publishers that there should be a book for physical geographers to do what *Geography and Geographers: Anglo-American Human Geography since 1945* (Johnston, 1979, 1983) had done for human geography. It may not be so easy to provide a survey of physical geography because it has not been so obviously affected by several paradigms and indeed in 1981 it was contended that amongst physical geographers geomorphologists at least lack a general theory and are in a conceptual vacuum (St Onge, 1981). Most previous attempts to provide reviews of physical geography have been concerned with parts of the discipline. In geomorphology there are available the first two volumes of a monumental history of the study of landforms (Chorley, Dunn and Beckinsale, 1964; Chorley, Beckinsale and Dunn, 1973). This present book cannot aspire to achieve the scholarship and authoritative vision conveyed by Chorley, Dunn and Beckinsale.

Two other books have dealt with part of the field of physical geography. Also confined to geomorphology and with an emphasis upon the early development in Britain was an important book by Davies (1968) in which the focus was admirably conveyed by the title *The Earth in Decay: a History of British Geomorphology 1578–1878*. A more recent volume also concerned with geomorphology is *The Nature of Geomorphology* (Pitty, 1982) which began its life as the early chapters of a geomorphology textbook which were subsequently extracted with their references updated to provide a survey of the overall nature of geomorphology. There has long been a tendency for geomorphology to be dominant in physical geography, so much so that in revising for examination questions, students have often to be reminded that where physical geography appears in a question then this is not synonymous with geomorphology! Although recent human geography has stimulated rather more discussions of the nature of the discipline than has physical geography there are some books which have relevance to physical geography and physical geographers. *Explanation in Geography* (Harvey, 1969) is a work of very considerable significance and an important perspective is provided more recently by *Geography, Ideology and Social Concern* in which Stoddart (1981) distinguishes the history of a subject which is merely the chronology of events from the history of geographical ideas. When reviewing recent ideas in physical geography as attempted in the succeeding chapters it is inevitable that reference must also be made to other disciplines. Thus the history of hydrology (Biswas, 1970) can be traced prior to 600 BC and developments in the eighteenth and nineteenth centuries provided the foundations for much subsequent work in hydrology and for the interest which physical geographers have shown in this field. Hydrology is spread across several academic disciplines and demonstrates how the subject matter of physical geography can be embraced within other disciplines in different countries. This dispersion means that it is imperative that the student of physical geography must not be content with a knowledge of British or North American physical geography alone. Whereas the Anglo-American view has been particularly pertinent in human geography (Johnston, 1979), in physical geography there have been important stimuli from other realms and the evolution of other related disciplines has perhaps been more inextricably interwoven with physical geography than is the case with human geography.

Throughout this book, which endeavours to provide a view of recent ideas and approaches in physical geography, it has been necessary to avoid drawing rigid limits that would preclude some research of particular interest to physical geographers, on the one hand, but on the other to avoid spreading the net too widely and making it appear that physical geography has embraced research undertaken by scientists in other disciplines. This is one constraint upon a book of this kind and five further constraints are considered in this chapter and succeeded by an outline of the approach to the followed in the subsequent chapters.

Constraints to consider

Retrodiction and reconstruction of the emergence of past ideas is not as easy as it first seems for at least five reasons which should be remembered when reading the remaining chapters. These extend from personal viewpoint and experience to selection of material, the time dependence of ideas, the scientific context at a particular time, and the temptation to suggest that a consensus of opinion exists where this may not indeed have been the case.

Personal viewpoint and experience are important because both are influenced by training and by knowledge of research developments in different parts of the world. Personal experience has been emphasized by Lowenthal (1961) in his contribution to behavioural geography where he argues that the world of experience of an individual is very parochial and covers a small fraction of the total available. Similarly in the research and writings of physical geographers it is possible to identify the way in which experience of particular environments, training in specific disciplines, or exposure to contemporary developments may all condition the individual perception of physical geography. The influence of environment upon the physical geography researched by physical geographers has been a recurrent theme for more than 100 years. Many illustrations are possible, e.g. the landscape features and deposits of Poland were major considerations in the development of periglacial geomorphology in that country in the 1950s and in the inception of the new journal *Biuletyn Peryglacjalny* first published in Lodz in 1954 with the leadership of the distinguished periglacial geomorphologist Jan Dylik. A further example is provided in a review of American field geomorphology where Graf (1984, p. 78) has suggested that spatial bias is a major hazard in geomorphic theory development because, with a small number of researchers, publications by a few individual scientists can affect the development of the subject. By analysing the geomorphology papers in 9 journals published in the US Graf (1984) showed how the 472 field localities for geomorphologic research papers 1817–1945 show a substantial spatial bias (Fig. 1.1A) and the 469 field localities 1946–80 (Fig. 1.1B) showed a spatial bias which was different and perhaps less pronounced. Graf concluded that modern American geomorphologic theory is likely to be spatially biased and that the classic humid cycle of geomorphologic theory was based on the assumption that many landscapes were similar to New England conditions or to those encountered in the western federal surveys. Furthermore, understanding of mountain landscapes and processes depends heavily upon observations reported from the Front Range of Colorado or the Sierra Nevada of California while approaches to the study of fluvial processes are prejudiced in favour of conditions in the Middle Atlantic states, the Rio Grande Valley of New Mexico, or western Colorado. It therefore seemed to Graf (1984, p. 82) that:

> The Southeast, South, and the Basin and Range Province outside California and Oregon remain the invisible regions of American geomorphology. . . . Field localities for future research opportunities might serve the science best if they are located in some of the geomorphologic blind spots.

Often related to this influence of environmental character has been the training and background experience of physical geographers. In some countries physical geography has naturally been allied with geology and with other earth sciences whereas elsewhere there have been closer links between physical and human geography. In Britain geology often tended to concentrate on the pre-Quaternary parts of the geological column

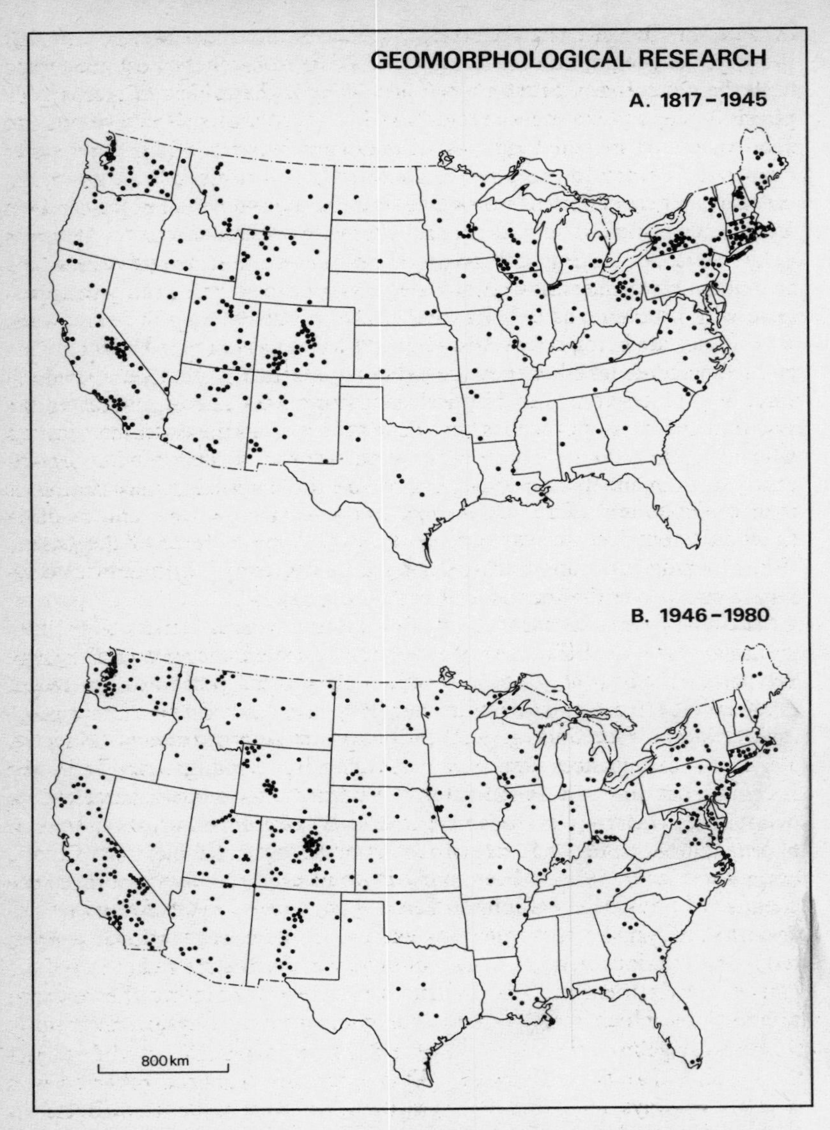

Figure 1.1: Field localities for major geomorphological research efforts as summarized by W.L. Graf (1984) for 1817–1945 in **A** and for 1946–1980 in **B**.

because the diversity and the research opportunities were so vast. This left the physical geographers for many years as the researchers most concerned with the Quaternary. Similarly palaeoecology is the subject of research by physical geographers in some countries but by botanists elsewhere and the subdivision of meteorology and climatology between disciplines varies from one country to another. Boundaries between disciplines are artificial and merely delimited for convenience and although some researchers would argue that one should not ask whether a particular scientist happens to be described as a physical geographer or not, it is necessary to know how individual researchers have been categorized in the past not only to appreciate why problems have been tackled in a particular way but also to see what topics are being researched now and will be investigated in the future and to speculate whether they are being approached in the most expedient way. Sometimes trends in physical geography have not responded to the way in which new approaches have been generated in the earth sciences as a whole. In the 1960s when environmental concern was growing significantly it is notable that physical geographers did not seize the opportunities then available and Hare (1969) noted how geography then and at other times had contrived to stay 'out of phase with the climate of the times'. When looking at the approaches in physical geography it is therefore necessary to range over the literature as widely as possible.

Selection of material has to be made from the writing known to an individual and this is difficult in view of the enormous growth in literature pertinent to physical geography which has been published in recent decades. However, the pace of growth may have now slackened and geography may now be finding itself in a steady-state environment (Haggett, 1977). As one glances over the products of physical geography in the twentieth century it is evident that the barrier effects of language and of delays in publication have been progressively reduced so that the influence of personal viewpoint and experience is declining. This has been further emphasized as there have been many more researchers contributing to the progress of physical geography so that it is no longer possible for one or two dominant physical geographers to exercise a paramount influence upon research directions in a particular country. But despite the reduction of the barrier effects there are still instances of research ignoring the research publications which already exist. For contemporary physical geography developed against a background of a dramatic expansion of the subject matter and of relevant literature. This expansion has been reflected in a number of ways including the growth in number of items included in abstracting serials, the development of new journals, and the output of books set against a background in which the number of societies was increasing and parallel trends and developments were affecting other subjects. Perhaps most significant has been the number of new journals introduced in recent decades and a number of titles and their dates of first publication are shown in Table 1.1. The recent growth of the contents of

Table 1.1: Some new journals of interest to physical geographers

Year	Primarily for geographers	Of interest to several disciplines
1947		Journal of Glaciology
1948		Geotechnique Arctic
1949		Journal of Soil Science
1950	Revue de Geomorphologie Dynamique	
1951		Bioscience
1952		Sedimentology
1954		Revue de Geologie Dynamique et Geographie Physique
		Biuletyn Peryglacjalny
1956		Hydrological Sciences Journal
1961		Water Resources Bulletin
1962		Journal of Hydrology (New Zealand)
		Reviews in Engineering Geology
		Journal of Climate and Applied Meteorology
1963		Journal of Hydrology
		Geophysical Research Bulletin
		Canadian Geotechnical Journal
1964		Canadian Journal of Earth Sciences
1965	Geografiska Annaler Series A	Water Resources Research
		Palaeogeography Palaeoclimatology and Palaeoecology Engineering Geology
1966		Earth Science Reviews
1967		Geoderma
		Journal of Engineering Geology
		Atmospheric Environment
1968		Quarterly Journal of Engineering Geology
1969		Arctic and Alpine Research
1970		Nordic Hydrology
		Quaternary Research Geoforum
1971		Water, Air and Soil Pollution
1972		Science of the Total Environment
		Ambio
1973		Catena
		Geology
		Coastal Zone Management
1974	Journal of Biogeography	Environmental Conservation
1975		Environmental Geology
1976		Geo Journal
1977	Earth Surface Processes (& Landforms)	Coastal Engineering
	Progress in Physical Geography	Polar Geography and Geology
1978		Journal of Arid Environments
1980		Annals of Glaciology
1981	Journal of Climatology	Soil Survey and Land Evaluation
	Applied Geography	
1982		Quaternary Science Reviews

Geo Abstracts as one of the major environmental abstracting journals which embrace physical geography is indicative also. Although this growth since 1960 partly represents the expansion and enterprise of K.M. Clayton and of Geo Abstracts, it also subsumes an increase in the available literature and in turn signifies the greater availability of the international literature to physical geographers. Whereas 50 years ago a single paradigm such as evolution could exercise a very clear direct and indirect influence, more recently it has seldom been the case that a single influence has prevailed to exclude the continuation of others that had been established earlier.

Time-dependence of ideas provides a third consideration in that in writing about past developments one is writing from the position of experience of present knowledge. In the case of physical geography it is not easy to reconstruct past thought simply in terms of the conceptual ideas available at that time. An example of this problem is provided by the drainage basin which is now accepted as a fundamental unit for analysis in hydrology and hydrogeomorphology. This unit had long been used as a fundamental unit before a paper on 'The drainage basin as the fundamental geomorphic unit' was written in 1969 (Chorley, 1969), but the origins of the concept (Gregory, 1976) are very difficult to trace. It was implicit in the calculations by Pierre Perrault of 1674 that the discharge of the river Seine was only some one-sixth of the rainfall received annually over the basin but it is difficult to establish exactly when the drainage basin concept became explicit. Therefore although it was not until 1969 that the significance of the drainage basin unit was summarized many implications had been assumed earlier and had been reflected in research but it is difficult to conjecture exactly what was appreciated in the past.

To emphasize the way in which any view, concept or explanation is really time-dependent it is desirable to utilize particular examples. Many explanations that are accepted may later be superseded or replaced and this is not simply what Carson (1971) referred to as the pendulum of fashion but is due to the fact that enormous leaps can be taken in scientific explanation in physical geography. The channelled scabland of Washington was suggested by Bretz (1923) to be the result of catastrophic flooding after the drainage of Late Pleistocene proglacial lakes that were released and produced flood flow which filled the pre-existing valleys to overflowing then produced divide crossings that converged to comprise an anastomosing complex of channel ways scoured in rock and loess. However, the features interpreted by Bretz, including giant current ripples up to 15m high, could not be accepted by the majority of workers in the 1920s and so Bretz encountered great opposition to his interpretation. This included the comments of W.C. Alden (1927 quoted by Baker, 1978c, p. 1252):

> It seems to me impossible that such part of the great ice fields as would have drained across the Columbia plateau could, under any probable conditions, have yielded so much water as is called for in short a time. . . .

Although the opposition was gradually overcome it was not until recent work by Baker (1978a, 1978b, 1981) that the enormity of the landscape development was appreciated, occasioned by maximum discharges which may have been as great as 21.3×10^6 $m^3.5^{-1}$ or 700 times the maximum discharge of the contemporary Amazon or 200, 000 times the maximum discharge of the Thames. This example illustrates how one assemblage of landscape features can be interpreted completely differently at distinct stages in the history of physical geography.

One expects to come across dramatic contrasts between nineteenth and mid-twentieth-century views but there are numerous cases where one particular feature or type of feature has been interpreted completely differently at slightly different times. This is illustrated by the features known as glacial drainage channels, meltwater channels or overflow channels. These channels, which are relict in the present landscape, often occupied anomalous positions on valley sides and often began and ended abruptly and did not have a fully satisfactory explanation until 1902. In that year P.F. Kendall produced a paper which subsequently became a classic (Kendall, 1902) in which he argued that such anomalous channels were the product of erosion by meltwater from proglacial lakes as the meltwater drained away either around the ice margin (ice marginal overflow channels) or through a col and over a watershed away from the glacier (direct overflow channels). These overflow channels were one line of evidence and, together with deltas produced where channels flowed into proglacial lakes, shorelines produced around the lake margins, and laminated lake deposits on the former lake bed, were taken to be indicative of the existence of former large proglacial lakes. This interpretation stimulated a generation of studies in many other parts of the British Isles (e.g. Charlesworth, 1929) and in other areas of the world as well. Very often because of the absence of other types of evidence the reconstruction of very complicated systems of freely draining proglacial lakes was proposed largely on the basis of the morphological character of overflow channels despite the lack of supporting evidence for the occurrence of such lakes. In 1945 a further classic paper was produced in Sweden (Mannerfelt, 1945) which demonstrated that 'overflow channels' could be produced by water flowing subglacially. By reference to contemporary glacier drainage he showed how a range of types of channel could be produced by water flowing in, under and on, as well as around, glacier ice. He also demonstrated that during ice decay wastage would lead to lowering of the ice surface as well as the recession of the ice margin. These ideas then stimulated a revision of ideas about the interpretation of the pattern of deglaciation and particularly as a result of the work of J.B. Sissons (1958, 1960, 1961, 1967) a model of glacier wastage involving an integrated network of erosion and deposition of meltwater came to be accepted and this led to the revised interpretation of the deglaciation sequence in a number of areas including the North York Moors originally studied by Kendall (Gregory, 1965), and to a much greater reluctance to propose the existence of enormous Pleistocene ice-dammed lakes. However there is no doubt that this second stage may have also been taken too far and knowledge of catastrophic Pleistocene events together with shorelines such as the Parallel Roads of Glen Roy led Sissons (1977) later to show how in some areas ice-dammed lakes did exist and had drained catastrophically giving a sequence of features which could have been developed in a relatively short period of time. Thus one of the major channels originally identified in the North York Moors (Kendall, 1902) could have carried water from small ice-dammed lakes and from melting ice in less than 30 years (Gregory, 1977).

This example is outlined to show how one set of features, namely glacial drainage channels, were central to a sequence of ideas and explanations. They were not satisfactorily explained before 1902 when they were explained as overflow channels. After 1945 they were increasingly interpreted as the routes followed by meltwater as well as drainage from ice-dammed lakes and as having occurred in positions round, on, in and under the ice so that the fragments that remain in the present landscape are the last surviving portions of an integrated network of meltwater flow. Since 1977 the possibility of catastrophic drainage of proglacial lakes has been envisaged and it has been realized that not all systems of deglaciation had to be the same, so that a more flexible model was adopted. It is possible to learn at least three things from this example. First, that the diffusion of ideas takes place gradually and that the rate of acceptance of a new model or explanation may be very slow and be accepted at different rates in different countries and by different disciplines. Thus although Mannerfelt's paper was published in 1945 debates concerning proglacial lakes were continuing more than 10 years later (Embleton, 1961; Derbyshire, 1962; Embleton, 1964). Secondly, that in the evolution of alternative models and hypotheses it is often necessary for some research to proceed to an extreme position before the inadequacies of the model can be exposed so that an alternative can be devised. Thirdly, that the development of new ideas often depends exclusively upon the imaginative leap or innovation provided by one researcher and that this has often occurred as an existing and accepted model has been tried in a new area or against an improved knowledge of environmental processes. Thus once again the environment and experience of the researcher can exert an important influence upon the interpretation of environment. The fortunes of the interpretation of glacial drainage channels and of deglaciation invite comparison with views on scientific explanation as outlined in chapter 3 (p. 42). The way in which new ideas and interpretation provide changing fashions is a theme which has been explored by Clayton (1970).

It is important to recall that new contributions to knowledge pass through a succession of stages and Stoddart (1981) quotes T.H. Huxley's four stages of public opinion as stated by Bibby (1959, p. 77) to be:

1 *Just after publication:* The novelty is absurd and subversive of Religion and Morality. The propounder both fool and knave.
2 *20 years later:* The novelty is absolute Truth and will yield a full and satisfactory explanation of things in general – The propounder a man of sublime genius and perfect virtue.
3 *40 years later:* The novelty will not explain things in general after all and therefore is a wretched failure. The propounder a very ordinary person advertised by a clique.
4 *100 years later:* The novelty a mixture of truth and error. Explains as much as could reasonably be expected. The propounder worthy of all honour in spite of his share of human frailties, as one who has added to the permanent possessions of science.

Although Huxley's view of the stages experienced by any new contribution may not be acceptable to all nevertheless it is undeniable that the significance of contributions is time-dependent.

The scientific context is important because physical geography not only reflects the influence of other sciences, especially the earth sciences, upon its development and upon its content, but also that of trends within, and pressures from, human geography. Such influences are potentially very significant in several different ways. Not only is the philosophy of science important, but prevailing trends in adjacent disciplines lead to new initiatives being taken. In the nineteenth century Darwinian evolution has been shown by Stoddart (1966) to have exercised a substantive influence upon physical geography and evolution ideas could subsequently be traced through the cycle of erosion in geomorphology, the zonal soil in pedology and the vegetation community in biogeography (see p. 18). In the twentieth century the second law of thermodynamics has provided the stimulus for the import and development of the systems approach and this has potentially provided an approach (Chorley and Kennedy, 1971) which is capable of use in geography as a whole (Bennett and Chorley, 1978; Huggett, 1980) although perhaps not yet as completely incorporated into some parts of physical geography as into other disciplines such as biology and particularly ecology (Allen and Starr, 1982; Margalef, 1968).

Developments in other sciences are also important because of the differing affiliation and training of physical geographers. Although physical geography is still prominent in university Geography departments in Britain, Canada, Australia, and New Zealand it has had varying fortunes in university departments in the United States. In the 1950s they tended to develop primarily as departments of human geography, and the physical geography that was included was overtly concerned with environmental description including what was described as landform geography (Zakrzewska, 1967) – whereas environmental evolution was the domain of researchers usually found in geology departments. More recently, with the advent of studies of landscape processes there has been a revival of physical geographers in the university departments of North America but some of the well-known contributors to physical geography have had their professional affiliation in other departments. Thus L.B. Leopold is a Professor of Geology and Landscape Architecture at the University of California Berkeley; S.A. Schumm is a Professor of Geology in the School of Forestry and Natural Resources at Colorado State University in Boulder, M.G. Wolman is Professor of Environmental Engineering at Johns Hopkins University in Baltimore, and A.N. Strahler was Professor of Geology at Columbia University of New York. The publication *Orbis Geographicus* lists world geographers and their interests but it is notable that many active contributors to physical geography will not be included because they are located in other departments and disciplines.

Costa and Graf (1984) summarized the way in which geomorphology emerged as a science in the US in the late nineteenth century when geology and geography were closely related disciplines. Subsequently geography became dominated by physical geographers and geomorphologists who had been trained as geologists but by the late 1930s geomorphology declined in importance and from the 1950s to the early 1970s the majority of geomorphologists in the US were not located in geography departments. In the 1980s however (Costa and Graf 1984) geomorphology is no longer required of undergraduate geology majors in many US Universities and this decline has coincided with the stability of the subject in geography departments as it has assumed a strong position and expanded into areas of natural hazards and natural resource management. Costa and Graf (1984) show how the professional activities of geomorphologists are spread across several organizations.

The training of physical geographers should also be considered because during the earlier days of physical geography it was usual for university researchers and physical geography specialists to be trained in some other disciplines simply because geography degrees including physical geography were not available. Thus the climatologist A.A. Miller was initially a graduate in mathematics and S.W. Wooldridge took his first degree in geology and it was on the basis of research undertaken in geology that Wooldridge was elected as FRS in 1969. L.B. Leopold took his first degree in civil engineering and his doctorate in geology. More recently it is not unknown for physical geographers to take degrees in subjects other than geography and M.J. Kirkby was for example a graduate in mathematics.

Interaction with different academic subjects becomes even more evident with the growth of interdisciplinary teams and with more complex subfields of research endeavour. It is traditional in some countries for a distinction to exist between institutions of higher education including universities on the one hand and Research Institutes which are branches of the National Academy of Science on the other and in which research is the main activity, Thus in the countries of eastern Europe, the USSR, and in China this dichotomy exists and it is essential that one does not assume that all of the active practitioners of the discipline are to be found in institutes of higher education because in the countries named the Academy of Sciences is the organization which often has the most active and prestigious research posts. In all countries there is a separation of research between universities or similar institutions and research institutions but the relative significance differs very substantially from one country to another.

The scientific context at a particular time is necessary to appreciate the extent to which physical geography has been influenced by, or has sometimes appeared to remain oblivious of, developments in science. Physical geography has almost reluctantly acknowledged the approaches to science since the publication of *Explanation in Geography* by Harvey in 1969 but some import has now been registered as indicated in chapter 3 (p. 41). However, physical geographers often write apparently unaware of the debates in the philosophy of science. For example a review of experimental method in geomorphology (Church, 1984) is completely consistent with the ethos of research in physical geography in not mentioning the work of Popper, Kuhn and Lakatos.

Also significant are the numerous important contacts with other sciences which will be evident in subsequent chapters and in fields like climatology, sedimentology, biogeography, process geomorphology and Quaternary geology, physical geographers now contribute alongside scientists from other disciplines. Because physical geography developed more recently than disciplines such as geology and biology there was a tendency to utilize developments in other fields and 'of all scholars geographers are most notorious to glean if not to harvest in other men's fields' (Spate 1960).

Much progress has been made in relation to other sciences but physical geography doesn't at present have a Fellow of the Royal Society amongst its ranks although there are a number of Fellows in adjacent fields of geology and ecology that overlap with physical geography. Because scientific enquiry is a continuum it is completely artificial to portray physical geography as a discrete and clearly defined endeavour. However it is equally unsatisfactory to resort to the notion that physical geography is what physical geographers do because without a decadal stocktaking there is a danger that certain tasks could be ignored or at least underestimated and this has been exemplified several times in the development of physical geography. What is required however is a point of view and this book endeavours to show how that point of view has developed, what its characteristics are, and to identify the major paradigms that have affected the emergence of the point of view.

Does a consensus of opinion and a single point of view exist however? It is inevitable that many different books could be written with the title of the present one. A point of view presented by one physical geographer is one of several and it must be remembered that each point of view is time-dependent. In the context of geomorphology, Yatsu (1971) argued that 'many roads lead to Rome, however, and many approaches are necessary for the harmonious development of geomorphology' and this could equally apply to physical geography as a whole. There are a number of occasions when the same general idea has been formulated and proposed independently in several different countries. Therefore to identify the attitudes and concepts of physical geographers it is necessary to appreciate the origins of ideas and also to appreciate how the diffusion of ideas takes place. The speed of diffusion of ideas has increased dramatically in recent decades and it becomes increasingly difficult for independent and distinctive schools of thought to persist as was the case in the earlier decades of the twentieth century.

One view of physical geography (Gregory, 1978a) visualizes the subjects of attention as expressed in a physical geography equation which embraces morphological elements, or results of the physical environment (F), processes operating in the physical environment (P) and the materials (M) upon which the processes operate over periods of time t. The equation thus can be expressed as $F = f(P,M)\,dt$ and it has been further suggested that studies by physical geographers can be envisaged as taking place at four levels of this equation and these are:

1 study of the elements or components of the equation – the study of surface form, climatic character, processes or materials in their own right. This is often a descriptive phase which can of course be quantitative, can embrace a considerable amount of innovation in the development of techniques, and is often preparatory to other levels.
2 study of the way in which the equation balances at different scales and in different subdivisions of physical geography. This could involve at the continental level demonstrating how the energy balance operates by relating available energy for environmental processes to radiation and moisture received in relation to materials locally available. At this level therefore the focus is upon equilibrium situations.
3 analysis of the way in which the equation varies over time and the way in which one equilibrium situation is disrupted and eventually replaced by another equilibrium. This stage can be thought of as differentiating the equation in mathematical terms and this type of study relies upon reconciliation of data obtained from different time scales and upon appreciation of some theory of adjustment of environments over time. At this stage of course the significance of human activity has to be appreciated because this is often the regulator that has altered an environmental system and has created a control system (chapter 7, p. 146).
4 application of the results of study of the equation depends very often upon extrapolation of past trends at spatial or temporal scales to locations for which estimates need to be made. This type of development can effectively only occur as a fourth stage and in the past was often not developed because of a reticence on the part of physical geographers

which prevented them taking their research to the logical conclusion of application of the results to contemporary and future environmental problems. This reticence has been overcome as physical geographers have become more involved in applied problems as shown in chapter 9 (p. 186).

Method of approach

Physical geography as described in this book may therefore be a prelude to a greatly changed physical geography by the year 2000 but it should be helpful to know of developments in the past so that one can more clearly establish the objectives for the future.

1984 is a good year to be finishing this book because physical geography is at an exciting stage for at least two groups of reasons. First because of the types of books and articles which are being published. These indicate how physical geographers are directing their attention more definitively to new subjects such as the urban environment (Douglas, 1983) and to the need for application of the results of research for example to the urban geomorphology of dry lands (Cooke, Brunsden, Doornkamp and Jones, 1982) and also are contributing in interdisciplinary endeavour, producing and editing works to which interdisciplinary scientists contribute. This includes dynamic meteorology (Atkinson, 1981), the environment in prehistory (Simmons and Tooley, 1981) changes in the water balance (Street Perrott, Beran and Ratcliffe, 1983), palaeohydrology (Gregory, 1983), modern and ancient fluvial systems (Collinson and Lewin, 1983) and hydrological changes (Walling, 1982). A further example of stimulating work of physical geographers is exemplified by *The Time of Darkness* (Blong, 1982) in which 54 versions of local legends in Papua New Guinea are compared with the scientifically reconstructed Tibito tephra eruption which involved a thermal energy production of 10^{25} ergs and so qualifies as one of the greatest eruptions of the last 1000 years. Simultaneously new technology is promising great strides in acquisition and analysis of data on the physical environment. The microchip and the microcomputer offer enormous potential in the extent and speed of data analysis and also in data acquisition as electronic wizardry is harnessed for further field monitoring. Remote sensing, which has been providing inputs to physical geography since the late 1960s, is now poised to offer even greater insights into the environment as new generations of satellite offer greater resolution and now begin to compete with field survey, by affording repeated coverage every 18 days or so, which field survey cannot. Greater use can also be made of the extent of the electromagnetic spectrum with radar rapidly giving information on the physical environment never available before. This coincides with a time of greater environmental awareness so that physical geography should be able to develop and respond to the opportunity available.

Although this is not a chronology of studies of physical geography it is inevitable that the approach adopted must be broadly chronological. Therefore the three parts are arranged in historical sequence, but to emphasize the most recent developments. Part 1 attempts to summarize developments before 1950, Part 2 suggests five approaches that have attracted attention particularly between 1950 and 1970 and Part 3 reviews two ways in which all approaches are now being further advanced. When the book was originally conceived it would have been appropriate to stop at 1980 but as it will not be published until the mid 1980s the final chapter attempts an assessment of the present balance sheet from which physical geography may progress to the next century.

Part I Antecedents 1850–1950

2

A century for a foundation 1851–1950

Justification for this chapter may be found in the fact that understanding physical geography is exactly like understanding environment: one needs some appreciation of how and why the present state developed. By 1850 there were established the clear beginnings of geography including physical geography. These origins were expressed in the foundation of geographical societies, and in the creation of university Chairs. The early nineteenth century was a time when many new scientific societies were being founded and the first Geographical Society to be founded was that in France, inaugurated in Paris in 1821, quickly followed by the German in Berlin in 1828 and the Royal Geographical Society in London in 1830. In addition to consolidating and concentrating effort in a growing discipline the new societies were created to form an avenue for publication. Sir Clements Markham, who was at one time a president of the Royal Geographical Society, pointed out in 1880 that, although the Royal Society had in theory published geographical work since 1662, only 77 of the 5336 papers published 1662–1880 could be called geographical. Each society formed tended to have its own distinctive focus; the Royal Geographical Society was very effectively involved in exploration but its journal was also the forum for many illuminating debates (Freeman, 1980). By 1866 there were 18 genuine geographical societies and by 1930 the number had risen to 137. University chairs of geography were also being established and in France there was a chair at the Sorbonne in Paris in 1809 and by 1899, chairs at five other French Universities (Harrison Church, 1951) and most of the more important German universities had a professorship in geography. In the United Kingdom the first Professor was Captain Alexander Maconochie at University College London, 1833–36 (Ward, 1960) and subsequently Halford Mackinder was appointed to the University of Oxford in 1887 and Yule Oldham to Cambridge in 1893. In the United States the first Professor of Geography was Arnold Guyot, who was appointed at Princeton in 1854 and before 1900 12 universities offered courses in geography, although not all had established geography as a permanent part of their curriculum, and physical geography was frequently taught in departments of geology (Tatham, 1951). As these chairs were established, the presence of a physical geographer could be very influential and, in Germany between 1905 and 1914, physiography was very prominent because of the influence of Penck in the prestigious chair at the University of Berlin (van Valkenburg, 1951).

Expansion did not take place uniformly over the century from 1850 but was subject to a number of controls which accelerated and retarded development in different countries. The details of trends in development need to be seen in relation to the influence of individuals or groups of individuals, and the foundation of the Institute of British Geographers in 1933 was a response by a group of people to a perceived need (Wise, 1983). The brief appearance of a *Journal of Geomorphology* (1938–1942) similarly was the result of the efforts of a number of enthusiasts including D.W. Johnson. A summary of the progress 1850–1950 is attempted first by considering the influences upon geography which were extrinsic and intrinsic and then attempting to consolidate the achievements of 1850–1945 and the attitudes and approaches of 1945–1950, which were to provide the foundation for major developments in the second part of the twentieth century.

Extrinsic influences

One can only really understand the development of physical geography in the context within which the subject developed and grew. This context was provided by extrinsic influences of two kinds: first those general ones which created the general scientific environment and therefore affected attitudes and approaches in physical geography, and second those specific factual and conceptual achievements that were realized by related subjects and which then provided some of the building blocks for research and investigation in physical geography.

Uniformitarianism

Perhaps the most persistent general influence upon physical geography and especially upon geomorphology, was the gradual acceptance of the *Theory of the Earth*, published as two volumes by James Hutton in 1795 and subsequently clarified by Playfair (1802) in his *Illustrations of the Huttonian Theory of the Earth*. This theory rejected catastrophic forces as the explanation for environment and gave rise to a school of Uniformitarianism whereby a continuing uniformity of existing processes was regarded as providing the key to an understanding of the history of the earth. The time necessary to overcome the catastrophists, and those such as Dean Buckland who believed that the world began in 4004 BC, provides a fascinating story which is treated comprehensively by Chorley, Dunn and Beckinsale (1964). They also demonstrated how Charles Lyell, who published a book on the *Principles of Geology* in 1830, and came to be regarded as the great high priest of uniformitarianism, assisted in achieving the acceptance of the Huttonian doctrine, although Lyell later did not subscribe to all implications of the Huttonian theory.

Uniformitarianism not only replaced the catastrophic ideas of landscape formation but also promulgated the idea that 'the present is the key to the past'. Although this was very satisfactory in that the earth's present surface provides the processes and mechanisms for understanding the past, it cannot be assumed that rates of operation of contemporary processes provide the key to the past. Undoubtedly the notion of uniformitarianism came to exercise an important influence in geology and then in physical geography, but more recently there were some suggestions that the doctrine may have been taken too far. Thus Sherlock (1922) was led to this conclusion as a result of his consideration of the extent of the effects of human activity by looking at man as a geological and also as a biological agent (see chapter 6, p. 118).

Evolution

In the late 1850s the attention of geologists and of potential physical geographers was distracted away from the implications of uniformitarianism. This was due to a further influence which clearly pervaded the whole of physical geography, namely Darwin's *Origin of Species*, published in 1859. Indeed the notion of evolution subsequently diffused from the biological into physical, social and mental spheres and in the first statement of the cycle of erosion in 1885 Davis termed his cycle a 'cycle of life'. Curiously, Hartshorne (1939, 1959) did not evaluate the impact of the theory of evolution on geography in detail, despite the fact that the book published by Darwin was probably the most influential book of the nineteenth century and one that had a tremendous impact on social and political thinking as well as upon geography. The impact upon geography has been resolved by Stoddart (1966) into four components. First was *the idea of change through time* and this is reflected in relation to evolutionary attitudes to the study of landforms following Darwin's own 1842 study of the evolution of coral islands and was particularly influential in relation to the cycle of erosion promulgated by Davis. Also in plant geography and in ecology a similar influence was being expressed and Clements, a man who held a position in plant ecology similar to that held by Davis in geomorphology, proposed plant succession as the 'universal process of formation development. . . . the life history of the climax formation' (Clements, 1916). The conceptual similarity of plant succession and the cycle of erosion was exposed by Cowles (1911) who produced 'physiographical ecology' by amalgamating Davisian geomorphology and Clementsian ecology

(Stoddart, 1966). Secondly, *the idea of organization* arose when interrelationships and connections between all living things and their environment found particular relevance amongst European workers who were concerned with community structures and functions and eventually with the idea of the ecosystem as expressed by Tansley (1935). This influence was particularly notable in attitudes to regions and in the early decades of the twentieth century the idea of organic unity temporarily served as a unifying theme. The third idea of *struggle and selection* and the fourth of *randomness and chance* did not have such a clear and immediate reflection in physical geography except that Darwinism was interpreted in a deterministic rather than in a probabilistic sense. The unique contribution of Darwin's theory, that of random variation, was neglected in geographical circles (Stoddart, 1966) and did not really appear in work by physical geographers until the 1960s. The effect of evolution was to impose upon physical geography a historical perspective that was to become a predominant influence upon geomorphology, upon the study of soils and of biogeography, and also found parallels in the study of climatology for at least 100 years. Perhaps it was the combined force of uniformitarianism and evolution that encouraged some of the more obscure manifestations of the historical approach.

Exploration and survey

A third general extrinsic influence occurred because the earth's environment was still the subject of exploration in the nineteenth century and even well into the twentieth. Therefore the results of expeditions and explorations were often the basis for new information which could then be incorporated into physical geography. In his review of the first 150 years of the Royal Geographical Society Freeman (1980) suggests that the initial purposes prominent in the society were exploration and map-making. It is important to recall that exploration not only provided data about areas hitherto unknown but also furnished additional data thus emphasizing that the description of environment is time-dependent. Thus it was possible to reconstruct what was known of the country back of Bourke in Australia with different levels of information detail returned in 1866 and 1956 (Heathcote, 1965). Exploration was important in stimulating progress in particular fields and the study of coral islands was initiated early because of travel by Charles Darwin (1842) and by J.D. Dana (1853 and 1872).

It was through travel to a number of areas that the glacial theory became established and the idea that glacier ice rather than diluvialism or icebergs could explain features and deposits of glaciated landscape, and L. Agassiz and De Charpentier together paved the way for the glacial theory. Very often exploration could provide familiarity with a new environment and so stimulate fresh ideas and opinions. In this way the explorers of the American West including John Wesley Powell, G.K. Gilbert and C.E. Dutton were notable (Chorley, Dunn and Beckinsale, 1964) and provided material and ideas that later came to be incorporated into geomorphology, often via the cycle of erosion. It has been assessed (Chorley, Dunn and Beckinsale, 1964, p. 591) that 'not since the time of Lyell had a new body of thought had such an immediate effect on geomorphic thinking in general. . . . particularly true in the United States. . . . Most landscape studies began to admit readily the power of sub-aerial erosion.'

Developments in surveying also provided an increasingly significant data source for physical geography. In Britain the foundation of the Ordnance Survey in 1795 and of the Geological Survey in 1801 were the beginnings of surveys which provided basic data and documentation for physical geography to proceed. In the course of geological surveying a number of important concepts were developed and the Geological Survey memoirs still remain as adequate testimony to the perceptive ability of early geologists. Other aspects of physical environment have been mapped or monitored much more recently and in the UK national soil survey dates from 1949, and the only vegetation surveys were embraced within the first (1930s) and second (1960s) land use surveys. Also in Britain environmental monitoring includes permanent records at tide gauges at coastal sites since 1860, of rainfall records since 1677 at Burnley, and of continuous discharge measurements on the Thames since 1883 but not until 1935–36 were there 28 river gauging stations in Britain and by 1945–53 there were 81. This number then increased dramatically so that by 1975 there were *c.* 1200 gauging stations. Approximately half of the area of continental USA was covered by soil maps 1899–1935 (Barnes, 1954) and by 1950 the US Weather Bureau had 10,000 regular and cooperative stations measuring precipitation. Systematic and continuous measurements of streamflow began in 1900 and the basic network of gauging stations was established during the period 1910–40. By 1950 observation occurred regularly at about 6000 points. This growth in environmental monitoring emphasizes how recent the acquisition of data on physical environment has been and also therefore why the advent of new techniques of data capture (chapter 10) promises such exciting times for physical geography. The recent development of environmental monitoring is indicated because an automatic weather station was shown to the Royal Society by Robert Hooke in 1679 (Rodda, Downing and Law 1976) and the first records of river stage included those for the Elbe at Magdeburg 1727–1869 (Biswas, 1970).

Conservation

A concern which began in the mid nineteenth century but which had little effect on physical geography until the twentieth century and even then received limited attention was the concern for conservation of environment. The conservation movement (Mumford, 1931) is generally thought to have started with George Perkins Marsh (1864) in a book entitled *Man and Nature*. The book was written as a 'little volume showing that whereas others think that the earth made man, man in fact made the earth' and also in the words of Marsh 'to suggest the possibility and the importance of the restoration of disturbed harmonies and the material improvement of waste and exhausted regions; and, incidentally, to illustrate the doctrine, that man is, in both kind and degree, a power of a higher order than any of the other forms of animated life, which like him are nourished at the table of bounteous nature.' This book, which included *physical geography* in its subtitle, proved to have a great impact on the way in which men visualize and use the land (Lowenthal, 1965) but as shown later (chapter 6), it was many years before physical geography fully realized and utilized the lead given by Marsh.

In succeeding decades experience of the sensitivity of certain environments prompted investigations, organization and publication that also provided information on the effects of use of the environment. Thus some of the earliest details of the physical geography of China that were published in the west tended to focus upon the erosion taking place in the loess lands of the middle Huang He basin, and this was referred to in a book on the *Rape of the Earth* (Jacks and Whyte, 1939) although most of the examples in that volume were derived from Europe and the Mediterranean, North and South America, Africa and Australia and New Zealand. A series of other books was produced in the 1930s and subsequently in the United States – as a result of experience of soil erosion produced by importing old-world farming methods into a new-world environment – and describing the measures designed to deal with the erosion problems which arose (Bennett, 1938). It is a curious paradox that this major problem of soil erosion did not attract the attention of physical geographers to any great extent at the time because of the lack of interest in landscape processes. Also influential in the decades before 1950 were a series of US Department of Agriculture Yearbooks and these mines of factual information included *Soils and Men* (Bennett, 1938) and *Climate and Man*, (Kincer *et al.*, 1941).

Specific foundations

It is somewhat artificial to separate intrinsic influences from more external ones because, particularly in the nineteenth century, it was not always clear whether a particular scientist could categorize himself or herself a physical geographer or not. The boundary with other disciplines remains blurred because not only in the USA were many physical geographers to be found in geology departments, but also in Germany von Richthofen, who followed Humboldt and Ritter, had trained as a geologist and Ratzel came to geography from early training in geology, zoology and comparative anatomy. In Britain university geologists like A.E. Trueman (Swansea), O.T. Jones (Cambridge) J.K. Charlesworth (Belfast) and A. Wood (Aberystwyth) made contributions which influenced physical geographers and advanced geomorphology in particular. In addition the contributions from scientists in organizations like the Geological Survey, the Meteorological Office, and the Soil Survey of specific countries together supplemented by an important amateur interest, all tended to provide specific foundations for the advance of physical geography. Despite the problems of selection, it is necessary to identify some of the specific foundations which were adopted by physical geography during this period.

In *soil science* there was important progress made by the Russian school of soil science which was headed by V.V. Dokuchaev. The prevailing perception of the soil had previously been influenced by agricultural chemists such as J. von Liebig in Germany, who had evolved the so-called 'dustbin theory' which visualized soil in a static way as a closed system into which one could simply replace what was extracted by crop production. This notion of a layer of soil as a skin, almost independent of above and below, was superseded by a view of soil as the product of soil-forming factors and together with his students, particularly Sibirtsev (1860–99), Dokuchaev was responsible for the zonal theory of soils which was published in 1900. This zonal theory produced the idea of zonal soils affected predominantly by climate; intrazonal soils determined by other factors such as water, rock type, or topography; and azonal soils which were immature and had had insufficient time for development to one of the other two types. This conception of broad belts of soils which were zonal and related to broad patterns of climate and vegetation was a natural development against the background of the latitudinally arranged physical environments of the USSR. Other Russian soil scientists included K.D. Glinka (1867–1927), who distinguished those soils which were amenable to climatic influences as ectodynamomorphic and those endodynamomorphic ones which had inherent characteristics which enabled them to resist, or at least modify, the influence of outside factors.

There was a delay before the Russian ideas of soil formation diffused to western Europe and North America, where soil was still regarded in a geological way. Diffusion was retarded because the translation of K.D. Glinka's work by C.F. Marbut, who was then head of the US Soil Survey, was based upon a 1914 German edition of Glinka's book and not published until 1927. Marbut was then well placed to amalgamate Russian and American ideas and published a work on the soils of the United States in 1935 which reconciled the needs of a national classification with the field survey scale, and he also published a number of articles relating to soils in the *Annals of the Association of American Geographers*. The emphasis which the Russian school had placed on factors of soil formation was not fully utilized until 1941 when Jenny produced a book entitled *Factors of Soil Formation* and used a central notion that any soil character (S) was a function of climate (Cl), organisms (O), relief (R), parent material (P) and time (T) as well as other unspecified factors in the form:

S = (Cl, O, R, P, T. . . .)

In *geomorphology* influences naturally derived particularly from geology and here the impact of geophysical developments was particularly notable. The theory of continental drift stated by Wegener in 1924 intrigued physical geographers, although this together with mountain building and earthquake activity was not given great cognizance. Perhaps the other extrinsic influences upon geomorphology were derived directly from geology and especially from the progress of geological mapping and survey, from the elucidation of stratigraphy and from the way in which the later stages of earth history and particularly the Cainozoic were being established. In geomorphology many of the major influences in this period were realized through the cycle of erosion and Davisian geomorphology (p. 29).

Studies of the *atmosphere* received a great impetus from the output from the Meteorological Institute at Bergen and it has been suggested that few groups have ever dominated a scientific field so completely as the Bergen group (Hare, 1951a). Stimulated by the lack of information for the 1914–18 war zone and facilitated by the dense network of stations which had been established especially in Norway, this Bergen group essentially placed fronts on the weather map. The central themes of the Bergen school centred upon the realization that the atmosphere is composed of large bodies of fairly homogeneous air separated by gently sloping boundaries or frontal surfaces. The life history of a frontal wave cyclone was developed by J. Bjerknes, and Bergeron in 1928 and 1938 proposed a classification of airmasses which provided the foundations of dynamic climatology. The possible explanatory description of world climates in terms of airmass theory was not pursued, to the detriment of the physical geographer (Hare, 1951a). A number of climatic classifications were produced at this time including the Köppen classifications dating from 1900–36 and the approach towards a rational classification produced by Thornthwaite (1948). Also of significance in climatology was the progress towards airmass climatology (e.g. Lamb, 1950; Belasco, 1952) but the leads embodied in these approaches were not followed up by physical geographers until much later. Studies of the atmosphere had also been lifted to a different plane with the advent of the radiosonde, which meant that small radio transmitters sent up by balloon could illuminate the record up to 50,000 feet, so that upper air synoptic charts and the significance of upper air investigations began to assist the interpretation of understanding of synoptic situations.

Work on the water balance by Thornthwaite (1948) was also one of the important developments that was to influence the development of *hydrology*. One of the first books on hydrology was by Mead (1919). There had previously been little positive and coherent achievement in this area despite the existence of practical works like Beardmore's *Manual of*

Hydrology (Beardmore, 1851, 1862). Emphasis upon the water balance provided one important strand in twentieth-century development and in North America important foundations were provided by Horton (1932, 1933, 1945) and by Sherman (1932), who initiated unit hydrograph theory.

In *biogeography* a major influence was that of Clements (1916) through the proposal of plant succession. This dynamic approach to ecology embodied the concept of climax vegetation in which a plant community has passed through all phases of its succession and has reached a state of equilibrium with the climate of the region. The early ideas of Clements may have fostered a somewhat unsatisfactory analogy between the development of a plant community over time and the life-cycle of an organism (Harrison, 1980). The sequence of vegetation development was visualized by Clements as taking place along specific pathways and involving five phases: Phase 1: *Nudation*, the initial creation of a bare area; Phase 2: *Migration*, the arrival of plant seeds; Phase 3: *Ecesis*, the establishment of plant seeds; Phase 4: *Reaction*, competition between the established plants and the effects which they have on the local habitat; Phase 5: *Stabilization*, when populations of species reach a final condition in equilibrium with the local and regional habitat conditions. During this vegetational sequence there were a series of transitional stages called seres and the final equilibrium was the climax vegetation. Much attention was devoted to the concept of succession by biologists and then by biogeographers and a number of features have been emphasized, such as the distinction between primary succession, which is the succession on new bare areas, and secondary successions, which develop on areas which have previously supported vegetation and where human activity or natural hazards modified the former vegetation and initiated the secondary succession. Harrison (1980) has suggested that the initial view of succession has been followed more recently by one in which succession is viewed as a process which results from plant-by-plant replacement and where the patterns generated by this replacement process have routine statistical properties. Whittaker (1953) stressed the evolutionary significance of succession and visualized the climax as a pattern of species abundances which is locally constant but varies from place to place, and Harrison (1980) concludes that this view is accepted by many biogeographers. Whereas initially it was thought that the climax community was climatically controlled, it was subsequently accepted that that climax communities are relatively permanent and stable and are adjusted to a particular blend of environmental and biotic conditions.

Also influential was the separation of ecology into autoecology, dealing with the environmental relations of individual plants, and synecology concerned with the environmental relations of plant communities. The concept of the ecosystem developed in 1935 by A.G. Tansley (Tansley, 1935) was subsequently influential in relation to the systems approach (p. 140). Tansley viewed the ecosystem when first proposed as composed of two parts, namely the biome which was the whole complex of organisms including plants and animals living together as a sociological unit, and the habitat or physical environment. Tansley viewed all parts of an ecosystem as interacting factors which in a mature ecosystem are in approximate equilibrium. This concept was very significant because it could be applied at different scales and because it focused attention upon the organisms, the environment, the circulation of matter and energy and the structural organization of the system. It therefore provided a conceptual framework for biogeography, was applicable at any scale between the biosphere and the individual acorn, provided a standard basis in terms of energy equivalence to compare different communities, and it incorporated human activity as an integral component (Tivy, 1971).

Positivist background

Increasingly evident in the scientific world during this 100 year period was the influence of positivism. This was established as a concept by Auguste Comte during the 1830s in France and was conceived to supersede free speculation or systematic doubt as defined by Réné Descartes (1596–1650) and was characterized by Comte as the metaphysical principle (Holt-Jensen, 1981). Positivist approaches came to be the foundation of what became widely known as the scientific method and depended upon making empirical generalizations, statements of law-like character, which relate to phenomena that can be empirically recognized (Johnston, 1983b, p. 11). Because positivism depends upon the use of empirical generalizations it embraces the verification principle because it demands testing of the empirical hypotheses which are proposed, leading either to verification or falsification. The aim is to achieve general laws which are not specific to a given set of circumstances, and positivism was given an enormous impetus by the advent of Darwin's *Origin of Species* and of evolution.

In the 1920s a group of scientists known as logical positivists emerged in Vienna and extended the fundamental principles of positivism by arguing that formal logic and pure mathematics, as well as the evidence of the senses, provide knowledge and they therefore opposed all unverifiable phenomena. The verification principle was central to the work of logical positivists and subsequently this promoted further debate relevant to physical geography (p. 47). However, Comte and positivism do not feature in the history of the study of landforms (Chorley, Dunn and Beckinsale, 1964) probably reflecting the fact that the approach that became so fundamental in science was not used directly and explicitly by physical geographers and also the fact that much of the influence in geography may have come into the subject as a whole because it led to the determinist–possibilist debate (Harvey, 1969). This was because Comte believed that not only do natural sciences seek the laws of nature but also that scientific investigation of societies would discover the laws governing society.

Much of the development of science was based upon a methodology which required the development of hypotheses by models, theories or laws and then by testing of these hypotheses using empirical data, so that in 1953 the function of science was expressed by Braithwaite (1953, p. 1) as:

> to establish general laws covering the behaviour of the empirical events or objects with which the science in question is concerned, and thereby to enable us to connect together our knowledge of separately known events, and to make reliable predictions of events as yet unknown.

The positivist approach, although not really assimilated in physical geography until the late 1950s and 1960s was already under fire in the scientific world. However, the advent of quantum mechanics and quantum theory, although initiated in the 1920s, did not begin to have an analogous influence in physical geography for half a century.

Intrinsic influences

Within geography a number of influences developed during the century from 1850 to 1950 and these influences have been the subject of elucidation and speculation since 1950. According to Harvey (1969), Hartshorne (1959) had provided a view of the subject in which he concluded that the Kantian thesis had been used by Hettner to establish that geography together with history and certain other disciplines was an idiographic rather than a nomothetic science. The exceptionalist view in geography derived from Kant (1724–1804) and had led to geography being characterized as a point of view rather than a subject concerned with a distinct subject area, and also concerned with a unique collection of events or objects rather than with the development of generalizations about classes of events.

Whereas this exceptionalist view was implicit rather than explicit as an influence upon physical geography, one influence from geography as a whole was the regional ideal. This had led to physical geography research being focused upon specific areas and *Stream Sculpture on the Atlantic Slope* (Johnson, 1931), *Structure, Surface and Drainage in South East England* (Wooldridge and Linton, 1939), and the *Geomorphology of New Zealand* (Cotton, 1922) are just three illustrations of a focus that was very evident in physical geography in the first half of the twentieth century.

A second influence within the subject arose from the experience of field areas. This emerged because of the association of physical geographers with the exploration of particular parts of the world, and German physical geographers, for example, were involved in research in South Africa especially in the Kalahari, in East Africa, New Guinea and in the Tien Shan amongst other areas. Such overseas experience sometimes arose because of colonial contacts. French physical geographers were active in Romania as well as in the Alps and in central Europe. Field experience also shaped theories of, and emphases within, physical geography. Thus the erosional effectiveness of glaciers over much of Europe was established by L. Agassiz and the experience of a succession of field researchers including Playfair, Jukes, Ramsay, Geikie, in addition to geologists in the American West, had established what Davies (1969) described as the 'fluvial doctrine' which acknowledged the efficacy of running water in shaping the land. Although concepts as fundamental as the effectiveness of glacial erosion and of fluvial erosion were gradually accepted in the nineteenth century, the process of acceptance often took several decades of heated debate and in the case of glacial erosion one view was that glaciers could no more erode the landscape than custard could erode the custard dish!

Thirdly was an influence which remained from the training and experience of early physical geographers. Several instances have already been quoted where physical geographers were trained in one discipline and then moved into and developed physical geography. Thus Marion Newbiggin trained as a biologist and in North America C.F. Marbut (1863–1935), who wrote a number of papers in geographical periodicals and was President of the Association of American Geographers when he gave a presidential address on 'The rise, decline and revival of Malthusianism in relation to Geography and the character of soils', was primarily a soil scientist who had trained in geology and proceeded to make a great contribution in soil science.

It is difficult in the second part of the twentieth century to imagine how much the development of physical geography in the first half of the century depended upon the diffusion of ideas. Barriers to diffusion sometimes were provided by language as in the way the work of the Russian School of Soil Science, formulated in the late nineteenth century, did not reach the USA until 1925 (Marbut, 1925) and then through the medium of a translation of the German 1914 translation of work by K.D. Glinka. This barrier is also exemplified by the time taken for the diffusion of Penckian ideas; although the Penck model was originally published in 1924, the difficulty of the material and its publication in Germany meant that few real appreciations of its import were developed until a translation appeared in 1953 (Czech and Boswell, 1953). Sometimes exchange visits or guest lectures can be seen to have accelerated the diffusion of ideas, and the period when W.M. Davis was exchange Professor at the University of Berlin in 1908–9 did much to introduce his ideas into central Europe and also produced one of the clearest statements of those ideas (Davis, 1912). Similarly it has been argued (Brown and Waters, 1974) that the visit of H. Baulig to give lectures in London in 1935 and their subsequent publication (Baulig, 1935) triggered a prodigious flow of studies of preglacial and glacial sea levels in Britain which occupied many British physical geographers for at least 25 years.

Physical geography by 1945: the impact of W.M. Davis

From 1850 to 1945 there was a tendency for somewhat diffuse material lacking a coherent approach to be succeeded by evolutionary-based approaches, which were followed by consolidation of, reaction to or extension of such evolutionary emphases. Books published during this period indicate a number of trends which are significant in relation to later development. These included the dominance of geomorphology and the fact that sometimes geomorphology was seen as the physical basis of geography, as in the book by Wooldridge and Morgan in 1937 – although when this book appeared in a second edition in 1951 the title and subtitle had been reversed to produce *An Outline of Geomorphology: The Physical Basis of Geography* (Wooldridge and Morgan, 1951). Furthermore, within geomorphology there was a dominance of Davisian geomorphology and this is reflected in the surge of American textbooks which began to appear in the 1930s and 1940s. Perhaps one other evident trend is the way in which a broad approach to physical geography, often styled physiography in the nineteenth century, was succeeded in the twentieth by a separation of geomorphology, climatology and biogeography and this trend is shown by the definitions compiled in Table 2.1.

Table 2.1: Some definitions of Physical Geography 1850–1945

Date	Authors
1868	T.H. HUXLEY 'We need what, for want of a better name, I must call Physical Geography. It is a description of the earth, of its place and relation to other bodies; of its general strúcture, and of its great features – wind, tides, mountains, plains: of the chief forms of the vegetable and animal worlds, of the varieties of man.'
1870	MARY SOMERVILLE Physical geography is a description of the earth, the sea and the air, with their inhabitants animal and vegetable, so far as regards the distribution of these organized beings, and the causes of that distribution.'
1898	W.M. DAVIS 'The successful development of Geography, considered as the study of the earth in relation to man, must be founded on physical geography – or Physiography as it is coming to be called – the study of man's physical environment.'
1903	R.S. TARR and O.D. VON ENGELN 'Physical geography, which may be defined as the study of the physical features of the earth and their influence upon man.'
1915	P. LAKE 'The geographer . . . is concerned with the surface of the earth and not with its interior, and in general he has only to consider the atmosphere, the hydrosphere, and the visible features of the lithosphere, or in other words the air, the ocean and the land.'
1944	K. BRYAN '. . . what is included in physical geography. Obviously all the factors of environment except those induced by the presence of animal and plant life are 'physical'. Physical geography has a unity when viewed from the standpoint of the human geographer. From the stand-point of the 'physical geographer', it is a group of special sciences, each pursued for its own end.'

isolated and debated are summarized at the beginning of chapter 3 (p. 41).

Reasons for the continued acceptance of the ideas of the Davisian cycle for many years have been reviewed by Bishop (1980). He argued that the cycle is not a scientific theory on at least two grounds; first because it is irrefutable in relation to the concept of stage and secondly because the theory has been modified in an *ad hoc* manner as objections have been brought against it. The concept that underlies Popper's definition of science as including falsifiable hypotheses of high information content, means that Bishop (1980) concluded that the Davisian hypotheses could have been of more value if they had been expressed in such a way that they could be tested by falsification.

There were some physical geography viewpoints and approaches to geomorphology in existence before Davis, and T.H. Huxley's *Physiography* (1877) was important not least because it endeavoured to present an integrated view of the physical environment by physiography, defined as 'the study of the causal relationships of natural phenomena or a consideration of the "place in nature" of a particular district'.

In an excellent and very readable assessment of Huxley's book Stoddart (1975) analysed the enormous success of this work, which was described as 'one of the best books read for many a long day' and 'a real service to the human race' (quoted in Stoddart, 1975, p. 21). It was due partly to the way in which the book began with the London basin and succeeded in arguing from the local and familiar to the unfamiliar. Physiography in Huxley's sense was particularly appropriate for the expansion of popular education in the decades of rapid industrialization, population growth and social awareness in the wake of *The Origin of Species*, it held a dominant position in British education for nearly a quarter of a century before being partly displaced by science subjects in their own right in the schools, but it never gained a central position in university geography because of the influence exercised by geology. The new science of geomorphology

> although still called physiography in the United States, was supplied with a new unifying principle, the cycle of erosion, a new technique, the historical analysis of landforms, and a new field of regional analysis, while at the same time abandoning climate, oceanography, biogeography and the study of human geography to other disciplines (Stoddart, 1975, p. 32).

In France, Emm de Martonne had produced a *Traité de Géographie Physique* in 1909 which was a book used world wide, later translated into other languages and which ran to six editions. Davis's *Geographical Essays*, first published in 1909, collected together a sample of influential papers and was widely read.

Developments in geomorphology after Davis can be viewed as developments, alternatives and objections. *Developments* occurred in research and also in textbooks where a range of books such as those by Wooldridge and Morgan (1937) and by von Engeln (1942) were amongst the ones that continued the Davisian ideal. *Alternatives* arose because some approaches such as that of Walther Penck (1924) depended upon an essentially different basis for geomorphology but one which was not so readily understood and applied, one which took much longer to be widely available in the English-speaking world (Simons, 1962) and one which was not accompanied by such a great volume of publication as Davis's ideas. *Objections* began to develop because some geomorphologists began to see possibilities other than a purely Davisian approach. Thus Kirk Bryan followed Davis to a large degree but also took account of Penckian views and together with his students made contributions in the newly developing field of periglacial geomorphology and also in arid geomorphology. Peltier (1954) argued that in geomorphology in the USA a new era began about 1940 and this new era was heralded by descriptive investigations particularly focused upon mapping; by dynamic studies of fluvial, solution, marine, periglacial, aeolian and volcanic processes; and by applied studies. Although such a trend began to occur it was also very apparent in the USA that physical geographers turned more to human geography than had been the case previously and that geologists went in search of oil! (Peltier, 1954, p. 366). Perhaps the most striking outcome of the Davisian period was not only the emphasis upon the long-term landscape evolution which was stated by Wooldridge in 1951 (Woolbridge, 1951, p. 170):

> A significant development of geomorphology in the present century has been the attempt to link its record with that of stratigraphical geology. As a pedagogic device designed to assist the explanatory development of landforms the concept of the cycle of erosion can afford to deal in purely quantitative terms.

But there was also attention devoted to regional geomorphology. A number of cases have already been mentioned (Wooldridge and Linton, 1939: Johnson, 1931; Cotton, 1922) and to these should be added the physiography of the eastern and western United States (Fenneman, 1931, 1938) and, although conceived independently, *The Alps in the Ice Age* (Penck and Bruckner 1901–9). The three volumes of the last-named work were important not only because of the suggested chronology in Europe but also because the four-fold subdivision of the glacial sequence that was suggested became extremely influential in the interpretation of Pleistocene history in many other areas of the world.

The regional theme was also evident in other branches of physical geography and was often supported by contributions from other disciplines. An influential book on *Climatology* first appeared in 1931 (Miller, 1931) and dealt with elements and factors of climate, then with climatic classification, and some 60 per cent of the book was devoted to the climates of the continents. The meaning and scope of climatology was envisaged as '. . . the geographer is more usually concerned with . . . the translation of average figures into certain biological responses.' At a rather different scale, Vishers's (1944) *Climate of Indiana* included more than 300 maps and also clearly illustrated the emphasis upon spatial patterns.

In plant geography, in addition to ecological plant geography embracing plant succession, Kuchler (1954) identified approaches centred upon vegetation maps, upon the historical development of vegetation, and upon floristic plant geography which endeavoured to establish the area occupied by a particular species and to explain its extent and location. These subdivisions of plant geography obviously resemble the subdivisions in geomorphology at the time, and Kuchler (1954) categorized the first decades of the twentieth century as devoted to accumulating information and the subsequent trend to coordinating data but he doubted 'whether plant geography will ever have its Darwin'.

In the first half of the twentieth century physical geography had tended to subdivide into specialisms to replace the more integrated physiography of the nineteenth century. However, not all writers agreed about the content of physical geography. Davis had preferred the term physical geography to the term geomorphology and even in 1951 Wooldridge and East (1951) in their book on Spirit and Purpose headed a chapter 'physical geography and biogeography', so that it was not absolutely clear whether physical geography should include biogeography or not. In addition to plant geography, biogeography would include animal geography or zoogeography and, when recognizing the three approaches of regional, historical and ecological, Stuart (1954) noted that it is unfortunate that the aims and methods of zoogeography have never been clearly formulated and that

> 'if scholars trained in geography and thoroughly grounded in the methods of regional study were as adequately trained in systematic zoology and palaeontology, they could be in a position to render important service to zoogeography' (Stuart, 1954, p. 449).

The oceans were already a part of the earth's surface which were beginning to be ignored by physical geographers, due in part to the independent growth of the science of oceanography (e.g. Sverdrup, Johnson and Fleming, 1942) although Burke and Elliot (1954) advocated a new approach to the regions of the oceans despite describing most of the professional geographers of the United States as landlubbers. More papers on soils appeared in geographical journals and Mabut attempted to apply the Davisian concepts of youth, maturity and old age to soils (Barnes, 1954, p. 386) and Isaiah Bowman (1922), who was for a number of years a director of the American Geographical Society, attempted to achieve a synthesis of soils, climate and vegetation in his book on *Forest Physiography*. Some definitions of physical geography dating from this period are indicated in Table 2.2.

Table 2.2: Some concepts and terms involving integration of aspects of physical environment

Term	Definition	Source
SITE	'. . a site maybe defined as an area which appears for all practical purposes to provide throughout its extent similar local conditions as to climate, physiography, geology, soil . . .'	R. Bourne 1931
CATENA	'. . a regular repetition of a certain sequence of soil profiles in association with a certain topography'	R. Milne, 1935
FLATS AND SLOPES	'The ultimate units of relief are flats and slopes, small units of a surface that are visibly either the one thing or the other and are not subdivisible on the basis of form. These are the electrons and protons of which landscapes are built'	D.L. Linton 1951
MORPHOLOGICAL UNITS	'Nature offers two inescapable morphological units and two only; at the one extreme the undividable flat or slope, at the other the undivided continent.'	D.L. Linton 1951
LAND SYSTEM	'. . an area with a recurring pattern of topography, soils and vegetation'	C.S. Christian, and G.A. Stewart, 1953

In his succinct review of the development of theory in geomorphology Chorley (1978) identified seven phases of development of the subject and the first three, teleological, immanent and historical, had certainly appeared before 1945 and could be extended to physical geography as a whole. The teleological view existed before uniformitarianism and survived until the late decades of the nineteenth century, when it was replaced by immanent ideas which concentrated upon explaining landform characteristics in terms of their inherent characteristics and underlying rocks, and a similar approach can be discerned in soil science. Historical approaches included those based upon the Davisian cycle of erosion and the Penckian model and they find analogues in the treatment of plant succession and in the concept of the zonal soil. The fourth basis for theory proposed by Chorley (1978) is taxonomic and although this had become established in several ways prior to 1945, there were other developments imminent in the last five years to 1950.

Systematic growth 1945–1950

After 1945, at the end of the Second World War, there was a sudden increase in the number of geographers and expansion of universities, there was a growth in student numbers and in the courses provided, some new journals such as *Erdkunde* were initiated (1945), and around 1950 there were several very significant papers which promised to shape physical geography in the next decade at least. As well as the younger students many peopled demobbed from the armed forces now completed their degrees and provided an influx of expertise to university staff, but many of these new or returning graduates had gained experience in a variety of areas including weather forecasting, air photo interpretation, and terrain analysis so that they were able to contribute a pragmatic approach which had not been so evident before. In 1951 Hare (1951a) noted that

> Climatology in a university Geography department is now very likely to be taught by a competently trained meteorologist with a much deeper understanding of his field than was generally the case before the war,

and some new techniques such as the advent of radar also afforded new lines of investigation in physical geography. This five-year period may be seen as the end of one century of development and also the beginning of two subsequent decades when new models, methods and paradigms began to appear. Therefore an attempt will be made to summarize the important developments in the five-year period which arose within and beyond physical geography.

In Climatology the tendency was for mapping and for climatic classification to become intensified and the advent of an approach towards a rational classification of climate (Thornthwaite, 1948) complemented the classifications earlier available from Köppen and A.A. Miller. In 1954 Leighly concluded that since the 1920s much academic instruction in climatology had been focused on the classification of climates and maps of climatic regions based on the classifications used and this subject was the subject of an excellent review by Hare (1951b).

In climatology growing interest centred on the frequency of occurrence of events and this found analogy in the analysis of frequency of occurrence of floods, which had prompted Gumbel to develop an improved statistical treatment of extreme values. Also appearing outside physical geography was work on evaporation by Penman (1948) leading to a clearer appreciation of spatial variation (Penman, 1950) and accompanied by new books of hydrology (Linsley, Kohler and Paulhus, 1949) to supplement the edited volume which had been available for nearly a decade (Meinzer, 1942). Although an interest in water supply problems was growing (Meigs, 1954) amongst physical geographers, they were still inclined to ignore hydrology in the way in which the oceans had similarly been underestimated and one of the outstanding tasks for a geographer was seen by Meigs (1954) as

> the development of a system of categories of water features; and when such a classification has been established, the mapping of the distribution of the different forms of water throughout the United States must be undertaken

thus emulating the classification and mapping approach that was appearing in many other parts of physical geography.

In Geomorphology important developments were about to occur. Although Davisian geomorphology was upheld by some practitioners such as Wooldridge, who still viewed physical geography as a subject which rested upon specialist sciences:

> Physical geography is, in a sense, better organized than its human or social counterpart because it rests upon specialist sciences like geology and meteorology which had made great progress before the aims of modern geography were formulated in any detail. There is thus no dearth, but rather an embarrassing wealth, of material out of which to construct the subject (Wooldridge and East, 1951).

Such an exceptionalist view was the type seized upon by Tudor David in his attack on geography as a whole (David, 1958) It is also notable that this view seems not to have appreciated that physical geography represented more than 10 per cent of all papers written in the examinations of the department of Science and Art, no other subject remotely rivalled physical geography from 1868 to 1876 and in many schools it had become a vehicle for instruction in the sciences, although it was bound to decline as those subjects became established (Stoddart, 1975).

Developments were essentially of three kinds: extensions, alternatives and additions. The Davisian school which had stimulated studies of denudation chronology was pursued in greater detail and led to *extensions* because the techniques were defined more precisely, for example, mapping and the analysis of long profiles (Brown, 1952) and the debates involved, for example that between a marine and subaerial origin for erosion surfaces (Balchin, 1952). In the discussion following this paper a diversity of views emerged including that of the geologist O.T. Jones (1952) who commented that:

> The idea that there are represented in this area, and elsewhere, a series of platforms due to marine erosion cutting cliffs back had become quite fashionable in recent years – much too fashionable. It seemed to have started with a visit to this country by Professor Baulig many years ago. Professor Jones had been asked to preside over that very provocative meeting, but had he known what mischief Professor Baulig would cause he did not think he would have consented to preside. Baulig may have been right, he may have been absolutely wrong; his argument however had been carried to a degree which would baffle the credibility of even the most naive person.

Although a wide variety of views had begun to appear concerning the nature of landscape evolution, these views could still be regarded as extensions of the Davisian approach, which was essentially historical in character. Alternatives to the Davisian approach appeared because of mounting criticism of the supremacy of Davisian geomorphology within geomorphology and in physical geography as a whole. This criticism (see p. 41) was particularly evident in the United States and Strahler (1950a, p. 209) commented:

> Davis's treatment appealed then, as it does now, to persons who have had little training in basic physical sciences, but who like scenery and outdoor life. As a cultural pursuit, Davis's method of analysis of landscapes is excellent; as a part of

> the basis for the understanding of human geography it is entirely adequate. As a branch of natural science it seems superficial and inadequate;

and in a retrospective assessment it was suggested that 'to some the Davisian method had become a stranglehold or at least a sedative' (Chorley, Beckinsale and Dunn, 1973, p. 753).

The *alternatives* arose because firstly, other models of landscape evolution were being considered and the results of early work on slopes owed something to the ideas of Penck (1924) which had been followed by Wood (1942) and were becoming incorporated into the work of L.C. King, who was already working on arid and semi-arid landscapes of Africa and in 1953 presented 'Canons of landscape evolution' (King, 1953). In the *Annals of the Association of American Geographers* for 1940 and 1950 were collections of papers which pointed the way towards alternatives, including approaches to slopes (Strahler 1950a) and a year earlier it had been argued that 'the geomorphologist may concern himself deeply with questions of structures, process and time, but the geographer wants specific information along the lines of what and where and how much' (Russell, 1949, pp. 3–4). Other alternatives to the Davisian approach began to arise by the study of other environments and particularly the significance of periglacial processes in past landscapes came to be appreciated following the work of Bryan (1946) and was developed into a periglacial cycle by Peltier (1950) although the diffusion of important work on the European continent (Dylik, 1952; Poser, 1947) into the English-speaking literature took until the early 1960s.

Additions to the Davisian approach came particularly from the field of glacial and coastal geomorphology. In glacial geomorphology, in addition to work on detailed areas by Pleistocene geologists, research on glacier fluctuations (Ahlmann, 1948) material embraced in reviews of *Glacial and Pleistocene Geology* (Flint, 1947) and on *The Pleistocene Period* (Zeuner, 1945) began to present a perspective of potential interest to physical geographers, although some, such as W.V. Lewis, had already established themselves in this field (Lewis, 1949). In studies of the Pleistocene new techniques were available (Zeuner, 1946) and hence this began to influence the study of biogeography, where pollen analysis in particular was to become a major technique for environmental reconstruction. The area of coastal geomorphology also was developing independently of Davisian ideas and this was exemplified by J.A. Steers's *Coastline of England and Wales* (1948).

The impression may be created that physical geography by 1950 was becoming a more disparate field of enquiry in which there were few signs of integrative study of the components of the physical environment, but some such signs did exist. Although mapping and classification of areas had become a feature of climatology, of geomorphology, of pedology and of plant geography in a number of areas all epitomized as taxonomic approaches by Chorley (1978), there were some attempts to approach the interrelations of environmental character, perhaps most noticeably in the USSR. There the influence of Dokuchaev and the school of soil science had prevailed so that in 1950 L.S. Berg had produced a volume dealing with the natural landscape zones of the USSR and with the interrelationships within the zones, and this was the foundation for later attempts to integrate physical geography (e.g. Suslov, 1961) and then for landscape science (see p. 199).

Such approaches had been attempted elsewhere by *Forest Physiography* (Bowman, 1922) and also at different scales. The small individual slope envisaged by Wooldridge (1932) as the atoms from which landscape was built was reconciled with information from other spatial scales in an influential paper on the delimitation of morphological regions (Linton, 1951) and this acknowledged the influence of the site (Bourne, 1931) and the catena (Milne, 1935) concepts. The two latter concepts were to influence the progress of soil survey investigations and the way in which soils surveyed in the field related to larger soil bodies and to other environmental characteristics. Two other trends towards more integrated approaches to physical environment were already evident. Land systems were introduced (Christian and Stewart, 1953) as a basis for the evaluation of land character in Australia by the CSIRO. Very different, but also potentially integrative in character, were the relationships between rainfall and runoff proposed by Langbein *et al.* (1949) against the background of analysis of the annual runoff over the United States, because this relationship was to be the stepping stone for later ones concerned with palaeohydrology and environmental change.

Some strands of possible integrated approaches to environment by physical geographers are intimated by the definitions cited in Table 2.2. Although such trends came at the end of a century and more of ideas, the prevailing emphasis throughout the century had been upon evolution of environment and the classification of environment. However the emphasis in evolution had been upon distant evolution so that the greatest concern had usually been with millions of years and upon the Tertiary, rather more than upon thousands of years and the Pleistocene, and few if any studies were directed to the last few hundred years. Similarly in classification, the focus was more upon static patterns than upon the interaction of components and upon the dynamics of environment. It was still usual to find the perception of physical geography as fulfilling an integrating role linking more specialist sciences together. Such an exceptionist viewpoint was difficult to sustain in 1950 but even more difficult to retain as the techniques multiplied and diversified in the next two decades. In those decades there are many ways in which the development of physical geography could be portrayed but the method adopted in Part 2 is to focus upon the themes of measurement (chapter 3), chronology (4), processes (5), man (6), and systems (7) each of which had at times been posed as the dominant, if not the only, paradigm prevailing for physical geographers.

In each chapter, selection of material must be ruthless but an attempt is made to capture some of the potential anticipated and realized in the period 1950–1970. When one looks at a sample of definitions of physical geography 1850–1945 (Table 2.2) it may be construed that physical geography was still in search of an identity. The next 5 chapters indicate five possible identities that have been considered particularly during the 20 years from 1950 to 1970. Definitions since 1945 (Table 2.3) emphasize a number of diverse viewpoints some of which have been alternatives and others which have occurred sequentially. Although some writers have discouraged the search for a precise definition (Leighly, 1954) and some have fostered an eclectic view of a subject which borrows its material from other earth sciences (Strahler, 1951), subsequent definitions have focused upon the relations with human geography, the integrity of the physical environment, the knowledge of physical processes, and the increasing potential for application of physical geography (Table 2.3).

Table 2.3: Some definitions of physical geography since 1945

Date	Author
1951	A.N. STRAHLER 'Physical geography is simply the descriptive study of a number of earth sciences which give us a general insight into the nature of man's environment. Far from being a distinct branch of science, physical geography is a body of basic principles of earth science selected with a view to including primarily the environmental influences that vary from place to place over the earth's surface.'
1955	J.B. LEIGHLY 'At the present juncture we should be better off without a sharply formulated definition. It would be good if we could again approach the earth with unhampered curiosity by whatever means the problems we encounter suggest. In particular we should discard a restriction that has long been laid upon us: the prohibition of concern with processes. Let processes be restored to the central position they deserve: physical processes in physical geography, historical processes in cultural geography'.
1962	F. AHNERT '. . . the place of physical geography seems identified clearly enough: it stands beside human geography and is closely connected with it by two-way causal relationships and common denominators of location, distribution, and areal differentiation. In this interdependence, physical geography serves human geography, but it also has a goal of its own: it views the physical earth not only as the stage of human affairs but also as an object of geographical research in itself, with an areal differentiation of phenomena that can be studied quite apart from its possible relevance to the existence and the works of man.'
1964	YU.K. EFFREMOV '. . . the subject of geography is the historically established and continuously developing landscape envelope of the earth, including all its constituent territorial complexes, the landscapes.'
1965	D.A. MILLER '. . . knowledge of physical processes at the surface of the earth adds depth to studies of man's modification of the earth for cultural and economic purposes, and how physical geography, sharing technology with specialized disciplines, benefits from being associated with the other surface-oriented aspects of geography.'
1971	N.A. GVOZDETSKIY, K.I. GERENCHUK, A.G. ISACHNKO and V.S. PREOBRAZHENSKIY 'Physical geography in the USSR has evolved as a separate discipline with

its own object of study and its tasks and internal structure; it is, to be specific, a discipline concerned with the geographical shell (the landscape sphere) of the earth and with its constituent natural geosystem (the geographical, or natural, territorial complexes). The methodology of the Soviet school of physical geography is firmly based on recognition of the objective existence of geosystems. . . . In such major Western countries as the United States and Britain, physical geography in the modern sense remains altogether undeveloped; it is being treated either as the sum of particular disciplines (geomorphology, climatology, and so forth) or as an auxiliary branch of geography that serves to introduce studies of socio-economic phenomena in their distribution over the surface of the earth.'

1975 E.H. BROWN '. . . physical geography is at the present time internally unbalanced, and centrifugal in character – to put it dogmatically, geomorphology plays too dominant a role in the subject. . . . Physical geography as an integrated subject has been rediscovered by non-geographers under the guise of environmental science, and . . . physical geographers have tended to hold aloof from this development.'

1978 R.J. PRICE '. . . Physical geographers should be concerned with the study of the totality of the natural environment and . . . the scales at which this study are undertaken must be clearly defined.'

1978 M. GRAY 'The science of geology has consisted of specialists, many of whom have done pure research which has later proved to be of relevance to man. To my mind physical geographers can play much the same role, in a discipline in which specialists research into worthwhile problems, pure or applied, often in collaboration with specialists in other disciplines. . . . I believe, therefore, that what the discipline needs is a majority of open-minded and rigorous specialists and an important minority of integrators.'

1980 A.R. ORME 'I prefer to define the field in terms of the physical processes and inorganic materials of our environment, recognizing a somewhat porous boundary with the biological processes and organic materials that are the principal focus of biogeography I am aware that others may include biogeography with physical geography. . . . I wish to proffer some advice to students and practitioners of physical geography. . . . Be prepared! Be integrative! Be predictive! And be Proud!'

Such definitions lead to the conclusion that physical geography was still in search of an identity. The next 5 chapters indicate identities that have been considered particularly during the 20 years from 1950 to 1970.

Part II Themes for two decades 1950–1970

3

Measurement mounting

This chapter considers not only measurement but also mapping, models, statistics and mathematics. Although these aspects could be viewed separately they do have common strands which were growing in the 1950s, contriving to produce an intellectual environment for physical geography in which measurement became the pervading feature of many new trends although it was also a hallmark of other investigations (see chapter 5). The topic of quantification is introduced by outlining the criticisms of preceding physical geographies, by sketching the scientific environment, by attempting a summary of the early, and then of the more specific developments in the branches of physical geography, and concluding with the achievements of two decades of quantification.

Declining Davisian didactics

The impact of Davisian geomorphology had been felt not only in geomorphology but also in other branches of physical geography which had often been constructed using the geomorphological trilogy of structure, process and stage and with the cycle of erosion in mind. Davis 'both organized and systematized geography in the United States and won recognition for the subject as a mature science and as an academic discipline' (Chorley, Beckinsale and Dunn, 1973, p. 734). In the development of the Davisian approach, not only had a means of explanatory description been provided to survive and develop (p. 29) but also a new terminology had been furnished with over 150 terms and phrases credited to Davis and probably at least a further 100 generated by his students. Peel (1967) concluded that:

> . . . the weight and vigour of modern criticism makes it perhaps surprising that Davis' views were accepted so widely and for so long without serious challenge

and also quoted, L. Lustig:

> . . . literally thousands of articles in the literature of all countries, produced by authors who believe that one need simply view an area, match it to some suitable block diagram, and then proceed to write a treatise on the history of that region.

Davis had acknowledged some new developments, and King and Schumm (1980), editing lecture notes for courses presented by Davis at the University of Texas in 1927 and at the University of California Berkeley in 1929, show how Davis had accepted parallel slope retreat and pediments. However, criticism was growing, much of it summarized in a special issue of the *Annals of the Association of American Geographers* in 1950. It is impossible to summarize all the criticisms of the Davisian approach but they may be thought of as concerned with detailed points, with emphasis within the theoretical model, and with the major focus that was offered. An attempt is made to summarize some criticisms under these three headings in Table 3.1. In a re-evaluation of the geomorphic system of Davis, Chorley (1965) has drawn attention to three major criticisms. First that it led to a dogma of progressive, irreversible and sequential change, signifying that the amount of energy for the transformation of landforms was simple and a direct function of relief or of angle of slope. This method of analysis of landscape was not supported when it was shown that, for example, drainage densities, erosional slopes and river meanders do not necessarily evolve cyclically. Secondly the emphasis was upon historical sequence rather than upon functional associations, which were more dependent upon process investigations (p. 87), and thirdly the approach was highly dialectical and semantic. As the volume of criticism accumulated, Wooldridge (1949) in his discussion of taking the Ge out of Geography expressed fear that 'the baby might be thrown away with the bathwater'. Indeed this may to some extent have occurred, producing a conceptual vacuum which had not been definitively occupied by a comparably broad systematic approach even as late as 1967. In that year Chorley (1967, p. 59) contended that:

> This deficiency has served to highlight many national preoccupations, some of long standing, with particular geomorphic objectives. Thus the development of the American style of 'dynamic-process' geomorphology, the Franco-German climatic geomorphology, the British denudation chronology/geological approach, the Polish Pleistocene-dominated geomorphology, the Russian applied geomorphology, the Swedish studies of process almost *per se*, the Eastern European morphological mapping, and the Central European tectonic bases have created a Godot-like atmosphere of articulate introspection.

It was in this conceptual vacuum or deficiency of an integrated approach to physical geography that quantification began to grow.

The atmosphere of science

The positivist stance in science had long assumed that measurement was a necessary feature of scientific investigations. An oft-quoted statement by the nineteenth-century physicist Lord Kelvin avowed:

> When you can measure what you are speaking about, and express it in numbers, you know something about it; but when you cannot measure it, when you cannot express it in numbers, your knowledge is of a meagre and unsatisfactory kind.

After 1950 there was an increasing attraction by some physical geographers to measurement motivated because new approaches depended upon measurement, because statistical methods were available for adoption, and because technology was making available data acquisition, processing and computing to handle larger amounts of information than could have been previously contemplated. It is tempting to suggest that physical geography was to some extent immune from dramatic impacts as a result of quantification because numerical data had always been employed, in climatology for example, but this generalization is a myth according to S. Gregory (1976):

> ..... another of our recent geographical myths, namely that physical geography has had less need of change in the light of the quantitative approach because it was already 'more scientific'. This may have been true in terms of its use of equipment, and it is also true that it could rapidly adjust from its previous qualitative model constructs of the Davisian era to the current quantitative models of our modern physical geography texts. But has there been the same adjustment, or even the awareness that adjustment was needed, in the light of the constraints imposed by, for example, sampling procedures, statistical testing of hypotheses, and the development of stochastic models? Somehow I doubt it.

It was not simply a matter of gradual adoption of quantification but of physical geographers taking note of advances in the philosophy of science. Whereas earlier physical geography had maintained a position without citing any philosophical considerations, such considerations now began to appear in the bibliographies of articles written by physical geographers. The position of physical geography could be pictured against the vision of science proclaimed by Kuhn (1962) and these ideas are a very evident influence upon *Models in Geography* as assembled by Chorley and Haggett (1967). Kuhn claimed that science is not a well regulated activity where each generation automatically builds upon the results of earlier workers. Instead it is a process of varying tension in which tranquil periods, characterized by a steady accretion of knowledge, are separated by crises which can lead to upheaval in the subject and to breaks in continuity. Paradigms were defined as 'universally recognized scientific achievements that for a time provide model problems and solutions to a community of practitioners'. The development of science takes place (Kuhn, 1962, 1970) in the series of phases in which a pre-paradigm phase characterized by conflicts focused around individuals, is succeeded by professionalization when definition of the subject is acute, and then by a series of paradigm phases each characterized by a dominating school of thought and each separated by a crisis phase when revolution occurs because problems accummulate which cannot be solved by the prevailing paradigm. Kuhn thus visualizes scientific activity as seeking solutions within generally accepted but often unspecified rules and conventions, and such puzzle-solving is characteristic of what he calls 'normal science'.

From such a viewpoint it is necessary to understand how science proceeds towards the construction of conceptual approaches which reflect the prevailing perception of the time, which is in turn regarded by Caws (1965) as a gross smoothing over of the surface of reality. In citing this work Harvey (1969) related the subjective pole of experience (S in Figure 3.1A) to sense perceptions or *percepts*, to mental constructs or images which provide *concepts* and to linguistic representation which gives *terms*. Harvey (1969) also reviewed the routes to scientific explanation including the Baconian route (Figure 3.1B) and the alternative which acknowledges dependence upon an *a priori* model (Figure 3.1C). Adoption of the first route (Figure 3.1B) can be dangerous because acceptance of the interpretations may depend upon the standing and charisma of the scholar involved (Moss, 1970). The second method begins with the researcher perceiving a pattern of some kind in the real world; an experiment is then formulated which can test the validity of the *a priori* model. The first route (Figure 3.1B) is effectively inductive by proceeding from unordered facts towards a generalization, whereas the second is deductive because it relies upon an *a priori* model which is perceived at an early stage to allow for manipulation of data and for conclusions to be drawn about some set of phenomena even if a complete theory is not available. An alternative *a posterion* model expresses the notions contained in the theory in a different form such as mathematical notation, but these have not been used as extensively in physical geography as have *a priori* models. Once a model has been tested by the route shown in Figure 3.1C this may lead to a theory, in Einstein's words: 'free creations of the human mind' but more specifically, a set of sentences, expressed in terms of a specific vocabulary, which will facilitate discussion of the facts which the theory is to explain. Whereas a theory cannot necessarily be shown to be true or false, a hypothesis is regarded as a proposition which can be shown to be true or false (Harvey, 1969) and is more restricted in science and Braithwaite (1960, p. 2) contends:

> A scientific hypothesis is a general proposition about all the things of a certain sort. It is an empirical proposition in the sense that it is testable by experience: experience is relevant to the question as to whether or not the hypothesis is true, i.e., as to whether or not it is a scientific law.

In the progress of science the aim is usually to establish scientific laws, which are defined by Braithwaite (1953, p. 12) as: 'equivalent to generalization of unrestricted range in space and time of greater or lesser degrees of complexity or generality'. It is difficult to countenance such restricted laws being extensively available in physical geography and therefore Harvey

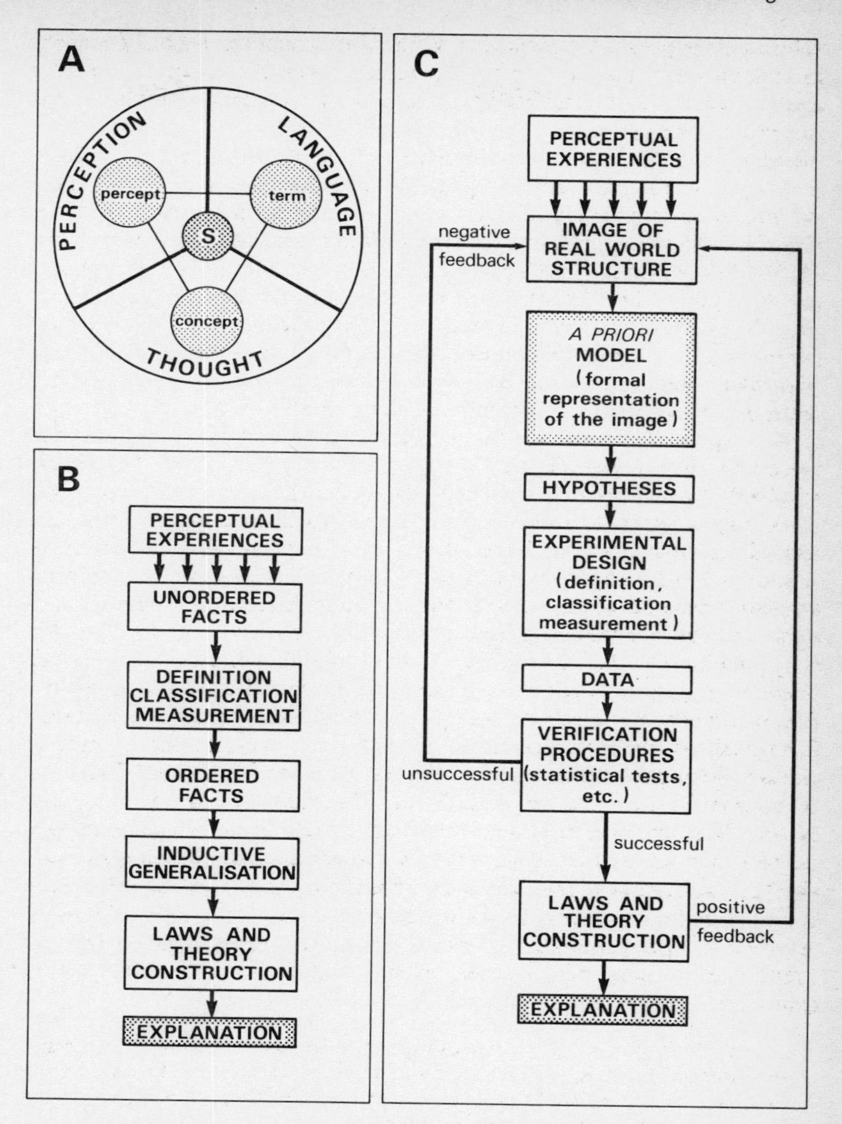

Figure 3.1: Patterns of scientific explanation (after Harvey, 1969). **A** provides a diagrammatic representation of the relationship between percepts, concepts and terms; **B** illustrates the 'Baconian' route to scientific explanation; and **C** indicates an alternative route to scientific explanation employing an *a priori* model.

(1969, p. 31) proposed a consistent hierarchy which could be particularly appropriate to geography. This hierarchy extended from lowest-order factual statements, to intermediate statements which could have the status of generalizations or empirical laws, to the highest-order statements which could be visualized as general or theoretical laws.

In order to proceed towards theories, hypotheses and laws, physical geography had to refer to the methods whereby scientific explanation is achieved (Figure 3.1B, C). With the advent of quantification the Baconian route was tried first, perhaps unwittingly, by some physical geographers in the hope that laws and theories would emerge from the whirlpool of numerical data. However, increasingly it was appreciated that models must be embraced within physical geography and from the late 1960s these were employed much more conscientiously in the (Figure 3.1C) route towards scientific explanation.

This route has occasioned an important debate which has been at least partially acknowledged by physical geographers. In scientific reasoning one of the central conditions is that theories must be supported by facts and several answers have been given to the question how exactly can facts support theory? In science generally Newton believed that he proved his laws from facts and before Einstein it was believed that 'Newton had deciphered God's ultimate laws by proving them from the facts' (Lakatos, 1978). Lakatos (1978, p. 2) continues to suggest that one can now easily demonstrate that there can be no valid derivation of a law of nature from any finite number of facts, so that if all scientific theories are equally unprovable 'what distinguishes scientific knowledge from ignorance, science from pseudoscience?' The positivist approaches underlying the routes to scientific explanation (Figure 3.1B, C) may be verified in several ways.

Paradigms were advocated by Kuhn (1962) as a means of visualizing the nature of science and during the paradigm phase the scientist accepts established theories, uses them as a framework for puzzle-solving and is not normally engaged in trying to overthrow theories and to develop new ones. Kuhn's work (1962) has been widely adopted in physical geography although it has been suggested (Masterman, 1970) that the different usages of the term paradigm by Kuhn could be categorized into three types:

1 a metaphysical way of viewing the world
2 a sociological notion which investigates how science operates
3 a 'classic' work which provides the tools and procedures whereby later puzzle-solving can take place.

The way in which a theory is verified is, according to Kuhn (1962), more a matter of faith than of logic, and the procedures of verification and confirmation are an integral part of the rules which the scientific community associates with the prevailing paradigm. Any anomalies that arise may then accumulate as the basis for the next crisis phase whereby revolution will engender the next paradigm phase.

An alternative school of thought has been generated, noting that no theory can be exposed to all possible relevant tests and by seeking criteria for evaluation of a theory in terms of the probability that the theory may be true in the light of the available evidence (Harvey, 1969, pp. 38–9). This required the production of an inductive logic of confirmation which would enable as objective a choice as possible to be made between alternative explanations. However, such a strategy was confronted by the difficulties of defining the relevant tests for a given hypothesis and of establishing the inductive rules to assess the degree of confirmation for a particular hypothesis. This inductive logic or probabilism was intended to define the probabilities of different theories according to the total available evidence. Thus a theory would qualify as scientific if its mathematical probability was high but if it was low or even zero then the theory was not deemed scientific. Thus probabilism offered a continuous scale from poor theories with low probability to good theories with high probability (Lakatos, 1978, p. 3). In 1934, however, K.R. Popper argued that the mathematical probability of all theories, given any amount of evidence, is zero so that scientific theories are not only equally unprovable but also equally improbable.

Critical rationalism has developed since the 1930s as a consequence of the works of K.R. Popper and is in opposition to logical positivism although D. Gregory (1978, p. 35) suggests that Popper had an extremely narrow conception of what positivism entails. The essential feature of Popper's view is that falsification replaces verification so that a theory is assumed to be true until it is shown to be false. In the course of scientific activity the scientist will propose trial solutions which are then evaluated critically and the trial solutions are speculative theories, set up in an attempt to solve the particular problem. Falsification as a procedure is justified because, whereas no finite number of facts can verify a universal proposition, a single fact can demonstrate the proposition to be false. Therefore scientific statements are conceived as being falsifiable whereas non-scientific ones are not. This critical rationalist stance of falsification has been criticized by Kuhn because if a single failure is the basis for theory rejection, then all theories ought to be rejected at all times. However if it is only severe failure which demands theory rejection then Kuhn suggests that the critical rationalists will require some criteria for 'improbability' or degree of falsification. Haines-Young and Petch (1980) contend that despite the lip-service paid to the critical rationalist position in the literature of physical geography, it has not been adopted as much as it might have been although it provided a more satisfactory basis for achieving further progress. They summarize three objections of the critical rationalist to the logical positivist thesis. First that facts are not objective because they are observations which are perceived in a particular way according to the technology available for measurement and observation and cannot therefore be shown to be true. Secondly that the principle of induction involves a logical error because no number of apparently confirming observations can show that a general proposition is true and thirdly that observations are not made independently of theory but rather that the variables perceived to be important and therefore selected for measurement are chosen in the light of knowledge of some preconceived theory. Haines-Young and Petch (1980) conclude that the discovery of previously unrecorded phenomena may arise from studies which fail to make their theoretical bases explicit and such discoveries can prompt generation of new problems and new theories; they also suggest that 'it is important to avoid massive empirical studies that are carried out with only the vaguest notion of a theoretical problem as if the accumulation of data was an end in itself' (Haines-Young and Petch, 1980, p. 75). It is unfortunate that they cite two examples of what they perceive to be such massive empirical studies without illustrating examples from their own research of work soundly based upon the foundation of a theoretical problem. This may be a danger of reviewing specific research against the epistemology or the theory of knowledge of physical geography.

In view of the dominance of Davisian ideas and their replacement (p. 41) it may at first appear that the Kuhn view of paradigm phases and of crises is appropriate to physical geography. Nevertheless the subsequent development of the subject with a number of paradigms prevailing as included in this and subsequent (4, 5, 6 and 7) chapters indicates that a Kuhnian view of normal science is not so readily applicable to the recent development of physical geography. An alternative vision was presented by Watkins (1970) when he suggested that a discipline can be in a multiparadigm state for considerable periods of time simply because it takes years to develop a new paradigm, and heretical thinking must have been going on for a long time before paradigm-change could occur. The contrast between one dominant paradigm – and the alternative multiparadigm – philosophy invites comparison with the cycle of normal erosion, which involved stability for long periods of time after a short phase of initial uplift, and with the alternative perceptions of landscape change which were to come later (p. 42). This multiparadigm line of development is embraced within the methodology of scientific research programmes advocated by Lakatos (1970) to solve some of the problems which both Popper and Kuhn failed to solve. Lakatos claims that the typical descriptive unit of great scientific achievements is not an isolated hypothesis but rather is a research programme, and that each research programme is supported by a heuristic, problem-solving machinery which, with mathematical techniques, can digest anomalies and convert them into positive evidence. All scientific research programmes may be characterized by their hard core of beliefs against which the thrust or *modus tollens* of the programme will not be directed because of the negative heuristic. The *modus tollens* is instead directed towards the protective belt of auxilliary hypotheses which is developed around the core and it is this belt that has to bear the brunt of tests and subsequently to become adjusted or even completely replaced. The positive heuristic is composed of a partially articulated set of suggestions on how to change or develop the refutable variants of the research programme and how to modify or sophisticate the protective belt. The positive heuristic therefore protects the scientist from becoming confused by an ocean of anomalies and will determine the problems which are chosen by scientists working in powerful research programmes.

Expressions of the quantitative revolution

It was long after 1950 that physical geographers gave some attention to the development of scientific theories and some would claim that even in the 1980s insufficient attention has been thrust in this direction. A quick check in the index of many recent books on physical geography for the names of Kuhn, Popper and Lakatos indicates how physical geography has still contrived to remain immune from developments occurring in science. Indeed when opening his chapter on bases for theory in geomorphology Chorley (1978, p. 1) wrote 'Whenever anyone mentions theory to a geomorphologist, he instinctively reaches for his soil auger' and, in relation to an all too prevalent view that there is no need to distinguish methodology from techniques and that the scientific method is obvious, commented (Chorley, 1978, p. 1):

> I do not agree with these views and, to adopt an aphorism, believe that the only true prisoners of theory are those who are unaware of it. It is indeed disturbing that attitudes to theory are so negative at a time when the bases of our theories are changing so rapidly and drastically with, I believe, profound implications for the future of geomorphology.

From clear beginnings in the 1950s the general environment of quantitative influences upon physical geography was created by trends within human geography and by general developments within physical geography. The development of the former has been ably chronicled by Johnston (1979, 1983a, p. 68), who has shown how the focus on theory and measurement together with the attempted development of 'geographical laws' was in keeping with the general ethos of academia in the post-war decades and accompanied the general movement towards a more nomothetic approach.

General geographic texts produced tended to reinforce and accentuate the quantitative trend and these texts may be perceived as being of two kinds. First there were general statements of the techniques that were available for adoption by all kinds of geographers and it is notable that two of these originated from the pen of physical geographers: one in an appendix (Barry, 1963) to an expanded version of an earlier book on cartographic techniques and a second as a text offering *Statistical Methods and the Geographer* (S. Gregory, 1963). Such works focused upon techniques for use by physical and human geographers and served to make known the statistical methods available. Emphasis was first placed upon such statistical techniques especially for hypothesis testing but was later supplemented by mathematical techniques. Introducing their book on mathematical methods Wilson and Kirkby (1974, p. 3) visualized progress over time towards deeper levels of understanding arising from the succession of orderly descriptions of the system of interest; to deeper understanding reflected in theories about why the structure of the system is of interest, how it works and how it changes; and thence to systems of interest which lend themselves to quantitative description which is often associated first with statistical and subsequently with mathematical analysis. It is perhaps paradoxical that many physical geographers have not fully realized the opportunities afforded by mathematics or acknowledged the differences which exist between mathematical and statistical methods, and the essentially descriptive quality of both groups of methods. Within physical geography researchers quarried the material presented for earth scientists as a whole and for geologists in particular, and prominent amongst these sources were Miller and Kahn (1962), Krumbein and Graybill (1965) and Davis (1973).

A second contribution to the general quantitative environment was provided by books which, although not exclusively devoted to physical geography, did present aspects of physical geography in a manner which contrasted with earlier approaches. *Frontiers in Geographical Teaching* in 1965 (Chorley and Haggett, 1965) was followed by *Models in Geography* (Chorley and Haggett, 1967), and by *Network Analysis in Geography* (Haggett and Chorley, 1969). In *Models in Geography* (Chorley and Haggett, 1967, p. 39) it was suggested that the traditional paradigmatic model of geography was largely classificatory and under severe stress, so they tentatively suggested an alternative model-based approach. Advocacy of this approach in 1967 was supported by the three minimum ingredients for the success of a new paradigm, namely that the new paradigm must be able to solve at least some of the problems that have brought the old one to crisis point; must appeal to the workers' sense of what is elegant, appropriate and simple; and it must contain more potential for expansion than the old paradigm which was being replaced. In *Network Analysis* inclusion of branching networks, which are tree-like in structure, and circuit networks, which are closed-loop structures, extended the focus of attention in terms of spatial structures, evaluation of structures, the relation to processes and location, and change by growth and transformation. This approach via networks, although very refreshing and original, was heavily dependent on certain parts of physical geography and particularly upon developments from R.E. Horton (see p. 56) and also itself had a structure which was too much at variance with the established subdivisions of physical geography into geomorphology, climatology and biogeography. However the book is an excellent exemplar of the way in which an intellectual environment was created by proposing a reorganization of geographical knowledge, by collating and sometimes importing material which would otherwise have been difficult to trace, and hence providing a stimulating and research-provoking work. A decade or so later similar collections of material spanning physical as well as human geography have been able to reflect the progress of quantification in two decades and include *Statistical Applications in the Spatial Sciences* (Wrigley, 1979) and *Quantitative Geography: A British View* (Wrigley and Bennett, 1981).

In addition to the intellectual environment created across geography as a whole and also extending over related parts of the earth sciences, there were some general review statements in physical geography. One very influential stimulus was the inception of a systems approach (Chorley, 1971) which led to an approach to physical geography as a whole (Chorley and Kennedy, 1971). In an article, published in 1971 in a journal which was born in 1969 in the aftermath of the quantitative revolution and was later (1977) to produce the offshoot *Progress in Physical Geography*, embracing excellent review papers charting recent progress, Chorley drew attention to the ever-deepening dilemma confronting physical geography. On the one hand physical geography had become responsible both for advanced research and for basic teaching in many earth science subjects and on the other hand it was required to play a relevant role in an increasingly economically and socially oriented human geography. Although both roles had been fulfilled efficiently in the past it was suggested that the increasing technical demands made upon researchers in the earth sciences, and the increasing preoccupation of human geographers with spatial socioeconomic matters was making the position of the physical geographer increasingly difficult, and this was pointedly characterized (Chorley, 1971, p. 89) as:

> . . . rather like that of a tightrope walker attempting to walk simultaneously on two ropes which are becoming more and more separated. Will the acrobat eventually commit himself wholly to one or the other of the ropes, thus either removing into the earth sciences the last vestiges of what used to be called the 'physical basis of geography', or continuing to feed forcibly increasingly unreceptive human geographers with a diet of what largely seems to them to be physical irrelevancies? On the other hand, will a painful schism occur rupturing traditional physical geography into two parts in which, for example, the 'geographical geomorphology' and descriptive climatology of 'true' physical geography are divorced from the geological geomorphology and dynamic climatology of the earth sciences?

In this penetrating analysis it was suggested (Chorley, 1971) that physical geography must be sensitive to the changing objectives of geography as a whole and also structured within a sufficiently viable intellectual framework. Three frameworks were offered based upon model building, which had been profitable but would not provide the kind of fusion needed between physical and human geography; upon resource use, which did not offer a sufficiently comprehensive intellectual base for academic physical geography; and upon systems analysis, which was the most attractive, was expanded in an important textbook (Chorley and Kennedy, 1971) and because of its centrality and development is the subject of a later chapter (7, p. 140).

Systems analysis was used as a framework for a summary of statistical applications in physical geography in 1977 (Unwin, 1977) because it was suggested that traditional subdivisions of physical geography were increasingly irrelevant to what physical geographers actually do. The major groups of statistical techniques discriminated were system description, the analysis of dependence, curve and surface fitting, the analysis of interdependence, and classification and discrimination related to morphological systems; stochastic processes, point and network processes, Markov processes, and time, distance and spatial series applied to cascading and process–response systems. This review (Unwin, 1977) clearly embraced the major groups of quantitative contribution but prompted the conclusion that greater emphasis had been placed upon deterministic than upon stochastic approaches; that quantitative advances and applications in 1970s did not continue the momentum generated in the 1960s; and that this loss in momentum contrasts starkly with the continued growth of statistical modelling in human geography. However writing in 1977 Unwin (1977, p. 208) suggested:

> . . . it may be that we are now on the threshold of a second wave of application of statistical methods to physical geography based not upon the classical statistics of Fisher and Pearson but upon methods developed originally for the analysis of fluctuations and noise in statistical physics where, like the environmental systems of interest in geography, 'chance' mechanisms play an obvious and important role.

In addition to the dominating emphasis upon statistical methods the increasing utility of computing methods (Mather, 1976), could highlight multivariate methods, and the general availability of mathematical methods (Sumner, 1978) provided two additional but diminutive props for quantification. Considerable scope still remains for the more extensive use of mathematical methods and Wilson and Kirkby (1974) concluded that:

> . . . there are many analytical problems associated with geographical systems of interest and many 'real world' planning problems, which could benefit from a more mathematical treatment

Themes in the branches of physical geography

Several features of the impact of quantification may already be evident and these include the way in which new statistical techniques were sought and imported into physical geography, sometimes with insufficient concern for data quality and problem specification where a technique was in search of a problem; the way in which batteries of techniques were presented to the geographic or physical geography audience sometimes with no real conceptual framework (e.g. Cole and King, 1968); and the manner in which conceptual progress by models, systems or networks was sometimes the vehicle for the selective presentation of techniques. Because the techniques often sought a problem within the established branches of physical geography it is to those branches that one should turn for the further illumination of the way in which progress was made.

The atmosphere

Study of the *atmosphere* was not exclusively or even predominantly the domain of physical geographers and Brown (1975) has drawn attention to the imbalance in physical geography and noted that, of the physical geography papers submitted to the 1972 International Geographical Union meeting in Montreal, 44 per cent were concerned with the land, 27 per cent with air, 9 per cent with water, 10 per cent with plants and animals and 10 per cent with soils. In Britain it has been argued (Unwin, 1981, p. 261) that:

> Climatology has always sat uneasily within British geography departments yet to date it has not been convincingly taken up by any other discipline. In consequence it has been depressed, a sort of Cinderella to the ugly sisters of geomorphology and biogeography.

and an inspection of nine English-language journals over the decade 1970–79 revealed on average less than one climatological article per journal per year (Atkinson, 1980). It was suggested that there were internal reasons embracing the expense and skills required, and external reasons, particularly the existence of government or national organizations, which accounted for this situation. However despite the somewhat obscure definition of the physical geographer's attitude to the atmosphere, it is in this field that internally and particularly externally the pertinence of quantitative methods became especially pronounced. The attitude to climatology as a form of book-keeping involving the numerical record of mean atmospheric conditions at particular places was attractive to many geographers and subsequently provided the basis for descriptive, statistical and physical climatologies that were effectively statistical in nature (Atkinson, 1980). By 1957 the distinction was clearly drawn between complex (analysis and presentation of climatic information related to practical applications) dynamic and synoptic climatology (Court, 1957). Synoptic climatology is devoted to obtaining insight into local or regional climates by examining the relationship of weather elements to atmospheric circulation processes. Dynamic climatology deals with the explanatory description of world climates in terms of the circulation or disturbance of the atmosphere (Hare, 1957). Climatology, which was now visualized as a subject for the geographer, is dependent on models in climatology which are primarily statistical in character, and also on meteorological models which are physical and mathematical and which rest on the basic laws of physics and hydrodynamics. Theoretical analyses of climate are almost exclusively undertaken in university and polytechnic departments of physics, geophysics, mathematics or meteorology or in government institutions such as the British Meteorological Office (Atkinson, 1980). Climatology therefore may be seen as the prime concern of the geographer and the contributions of geographers were resolved by Atkinson (1980) into four parts. First regional-physical, concerned with classification and particularly following the rational physically based approach to classification inspired by the work of Penman (1948) and Thornthwaite (1948) and also with heat and water budgets of global or continental areas. Secondly synoptic climatology concentrating upon data methods and applications to link global understanding of the atmosphere and knowledge of local and regional scale phenomena and well exemplified in the book by Barry and Perry (1973). Thirdly, boundary-layer climates which include the physical, topo-, local-, meso-, and regional climates of the lowest kilometre of the atmosphere, and embrace both natural and man-modified climates (Oke, 1978). Fourthly climatic change has also been an arena for work by physical geographers. In each of these four contributions, quantitative methods were a necessary tool and were utilized by physical geographers without the drama that precluded the adoption of such methods in geomorphology. Thus in studies of climatic change Bennett (1979) classified the methods used as based on eyeballing by intuitive methods, on extrapolation of the statistics of surrogate data, for example from data on frozen seas or rivers, and on black box approaches searching for trends and periodicities by statistical methods. Use of these three methods was confronted by difficulties arising from choice of variables, the range of variation encountered, the equation structure, sources of non-linearity and non-stationarity, the consequences of the observed record, feedback loops, confidence intervals, and the series length available for analysis. Although the focus of research on the atmosphere by geographers has centred upon climatology it has been argued (Hare, 1966) that as the shift occurred away from such parameters as temperature and humidity and towards the measurement of fluxes, the geographer should be exclusively concerned with climate as environment and therefore with the direct impact of climate on human health, efficiency and psychology and its indirect effect on economic activity (see p. 205).

The need for use of quantitative methods in climatology was further emphasized by the wealth of existing data and by the acceleration of data collection. To supplement the massive amounts of data already available from existing synoptic observing stations and from upper-air radiosonde stations, attempts to increase the quality and quantity of observations were made in the late 1950s. The World Weather Watch (WWW) was instituted in 1968 and for any 24-hour period transmits to processing centres standard meteorological observations from more than 9200 land stations making surface observations, from nearly 1000 stations making upper-air observations, nine fixed-ocean weather ships, together with data from 7400 merchant ships making surface observations only, and reconnaissance and commercial aircraft providing more than 3000 reports daily (Atkinson,

1980). The WWW led to research activities under a Global Atmospheric Research Programme (GARP) which aimed to provide a more fundamental understanding of atmospheric circulation and of the climate system as a whole. The development of the use of remotely sensed information from satellites also catalysed the increase of data availability and the initial GARP Global experiment (FGGE) was the first occasion on which observation for one year (1 December 1978 to 30 November 1979) included observations from a truly integrated system of satellites used to observe the earth's atmosphere. Five geostationary satellites continuously monitored the equatorial and sub-tropical belts and a series of polar-orbiting satellites were employed to determine the temperature structure of the atmosphere and to provide information on cloudiness and on sea temperatures.

Because of the wealth of data available, statistical methods are of primary concern to the physical geographer climatologist. In their textbook Barry and Perry (1973) devote some 14 per cent of the text to statistical methods organized in four sections: frequency and probability analysis; time series; spatial series; and classificatory methods which had not been fully explored in climatology but which offer potential in the light of experience gained by other disciplines.

However a further reason underlining the necessity for quantitative methods in climatology was the dependence upon fundamentals of meteorology which employed basic laws of physics and hydrodynamics and in which a number of dramatic developments occurred. In addition to developments in availability of observations there were great advances in the mathematical modelling of atmospheric circulations at several scales and it was meteorological models which dominated a review of models in meteorology and climatology (Barry, 1967). Seven fundamental principles were enunciated (Barry, 1967, pp. 97–8) as a basis for understanding a great range of atmospheric models namely:

1 The first law of thermodynamics – referring to the energy within a thermodynamic system and stating that the heat supplied to a gas is equal to the increase in its internal energy plus the work carried out in expansion against its surroundings,
2 the second law of thermodynamics – the entropy of a closed system either increases or remains constant during any process operating within the system;
3 the geostrophic wind equation which is a model for wind flow based on Newton's laws of motion and which expresses wind velocity as a balance between the horizontal pressure force and horizontal coriolis deflection due to the earth's rotation when there is no friction and no curvature of the isobars;
4 the thermal wind equation which relates the geostrophic wind at an upper level to the low-level geostrophic wind and the mean temperature in the intervening tropospheric layer;

5 the continuity equation which is the principle of conservation of mass;
6 pressure tendency – i.e. the rate of pressure change at sea level which depends on the vertically integrated net conveyance;
7 the vorticity equation where vorticity is a measure of the rotation of an infinitely small fluid element and relates to the conveyance.

The development of circulation models of the atmosphere was assisted by conceptual developments including the Rossby interpretation of a three-cell model and the significance of the jet stream and promoted two basic kinds of atmospheric models: barotrophic in which isobaric and constant density or constant temperature surfaces are parallel; and baroclinic in which the surfaces intersect. Of the several general circulation models developed in the 1960s and 1970s the Rand Corporation model is reviewed by Atkinson (1978) who shows how climates of a typical January and July have been simulated in terms of global distributions of monthly mean values of surface pressure and temperature, relative humidity at the 800 mb level, evaporation rate and precipitation rate. Despite a number of limitations, such as an over-vigorous hydrological cycle, the simulations are suprisingly good and the advent of such models has implications for the investigation of longer-term climatic change (see p. 166).

Geomorphology and the land surface

In geomorphology perhaps the most influential paper, triggering direct and indirect quantitative developments, was that by Horton in 1945. Although this paper was not the first in which Horton had developed concepts that were later to assume great significance, it was the initial comprehensive statement which embraced two major contributions. First it provided the foundation of the Horton runoff model (see p. 101) and hence stimulated the process revival, and secondly it offered the basis for a quantitative approach to land morphometry. For these two reasons it was probably the paper most cited in the research writings of physical geographers in the subsequent two or three decades. The attraction focused upon stream ordering and upon two 'laws' of drainage composition which were proposed and subsequently expanded to five. The attraction of Horton's approach derived from the facts that it was quantitative, it offered a way of relating form to process, and it appeared to be more closely related to contemporary landscape and its problems than did denudation chronology. The approach was adopted and developed by A.N. Strahler and by a succession of his students including S.A. Schumm, D.R. Coates, M.A. Melton, M.E. Morisawa together with V.C. Miller, J.C. Maxwell and A.J. Broscoe, who were to develop the Columbia University school of geomorphology and later to pioneer their own new approaches such as the contribution to the process school by Melton and others (p. 96). This was the first major school of geomorphologists to challenge the denudation chronology school in Britain and in the preface to *Network Analysis in*

Geography (Haggett and Chorley, 1969 p. v) it is acknowledged:

> If books are traceable through misty taxonomic trees to distant intellectual forebears, then this book might claim ancestry in Robert Horton's remarkable paper on the erosional development of streams. This 1945 paper by an American civil engineer laid the basis not only of much analytical work in fluvial geomorphology but more general underpinnings for a process approach to quantitative morphology. Its findings were taught to one of us by Vaughan Lewis at Cambridge and to the other by Arthur Strahler at Columbia; for both of us it represented the start of two trails of work, the one moving away from geomorphology into locational analysis and the other towards the study of drainage basins as the fundamental geomorphic unit.

The Strahler school at Columbia stimulated work on drainage basin morphometry and after a period of nearly a decade it was appreciated that stream ordering and drainage composition were reflecting statistical relationships rather than deterministic ones, but the bonanza of developments is reflected in the summary provided by Strahler (1964). Morphometry promoted several themes in subsequent quantitative research. Methods of ordering were necessary as one of the foundations for a more consistent comparison of basins anywhere in the world and together with a range of drainage basin characteristics provided a category of drainage basin parameters that were later to be utilized in relation to models relating indices of climate, drainage basin character and measures of discharge and sediment yield (Gregory and Walling, 1973). In addition there were studies of drainage networks which proceeded via the related approaches of simulation and topology. The analysis of network topology produced a substantial number of papers which placed more emphasis on the morphological properties of the network than upon the functional significance so that Werritty (1972 pp. 193–4) concluded retrospectively that:

> The mathematics had proved both elegant and exciting, but from the geomorphic viewpoint, exploration of the purely topologic properties of stream networks has proved unrewarding;

and Smart (1978), who was contributed so much to this type of quantitative research of TDCN (topologically distinct channel networks), concluded in an extensive review that network topology could not yet beneficially be utilized in relation to hydrological prediction models:

> The major difficulty in obtaining unambiguous quantitative evidence of the effects of network composition on hydrologic properties is that of disentangling them from the influence of a number of other complex phenomena.

Other developments in geomorphology were either conscious of the research inspired by Horton or had crystallized independently. In a singular approach Scheidegger (1961) devised a *Theoretical Geomorphology* which was expanded in 1970. Such a theoretical approach was conceived from the viewpoint of geodynamics and did not receive the acknowledgement it deserved because it relied upon a mathematical theoretical foundation and did not cover the range of geomorphology completely. However as an intriguing and stimulating approach it would probably have received more acclaim had it been launched up to decade later to coincide with the movement towards a more theoretical foundation and particularly towards the properties of material (p. 102).

Slopes also offered scope for a new approach to geomorphology and several origins could be detected. Some of the work of the Columbia school was focused upon slopes and two papers by Strahler (1950) focused geomorphological attention upon the need for slope analysis and in particular upon the use of frequency distribution analysis. In one of the later influerttial studies of the Columbia school Schumm (1956) undertook a classic study of badland type morphology on weak clays at Perth Amboy and successfully linked the Horton morphometry approach with the study of slope processes, producing drainage density values that held the world record for some time with ranges between 313 and 820 km km^{-2} for second-order basins. At least two other types of development characterized slope study and one orginated in the Netherlands where in a series of papers Bakker (e.g. Bakker and Le Heux, 1946) endeavoured to develop mathematical models and this was a task also undertaken by Scheidegger (1961). This theoretical approach did not find favour in some quarters and Doornkamp and King (1971, pp. 199) dismissed them because

> it is our contention that slope studies and development theories should be related to field measurement and not solely to theoretical abstraction.

Young (1963) produced deductive models of slope evolution but these were derived from the second origin, namely morphological mapping (Waters, 1958; Savigear, 1965). Not only did slopes require more investigation but methods of providing slope data had to be found, and morphological mapping offered an approach which had the advantage of concentrating on the entire landscape assemblage in contrast to the emphasis in denudation chronology where, as Young (1964) noted, there may be only 5–10 per cent of the landscape investigated. In some ways such morphological mapping encapsulates the developments taking place in the 1950s and 1960s because slopes were a new area being explored, data deficiency was being remedied by mapping methods but the enormous amount of data collected had to be analysed. Methods had to be sought for such analysis, data processing providing one alternative (Gregory and Brown, 1966) although this was still based rather more upon the approach in Figure 3.1B than that in Figure 3.1C! Elsewhere mapping was less dependent upon morphological data and geomorphological maps were produced for at least parts of a number of east European countries (p. 199). More recently the extension of measurement of land surface form has promoted a field of general geomorphometry defined (Evans, 1981, p. 31) as analysis of 'the land surface as a continuous, rough surface, described by attributes at a sample of points or of arbitrary areas'. Such general geomorphometry is claimed to avoid problems of landform definition and delimitation which are more apparent in specific geomorphometry which is based upon the definition of specific landforms and their quantitative description.

Sedimentology and coastal problems in particular also provided an additional inspiration for quantitative developments and this was directed particularly towards the description of sediment characteristics so that they could be the subsequent basis for deduction about environment and landscape change. Work on grain size characteristics of sediments involved the development of appropriate statistical techniques for description and comparison of sample frequency distributions and this type of advance is reflected in the content of quantitative texts produced (e.g. Krumbein and Graybill, 1965; Miller and Kahn, 1962). For glacial deposits the possibilities introduced by till fabric analysis (Holmes, 1941) were beginning to be appreciated and research in the 1960s demonstrated how quantitative methods could be employed (Andrews, 1970) for the analysis of the orientation of coarse particles in tills and similar methods were subsequently employed for fluvial deposits. A further aspect of coarse deposits which had attracted attention was particle shape and, although this could be assessed by simple visual estimation (e.g. Krumbein, 1941), quantitative measures of flattening (Cailleux, 1947) and of roundness (Tricart and Schaeffer, 1950) were the basis for environmental interpretations from specific deposits and later for more comprehensive proposals such as the 13 partially independent parameters needed to characterize a particle or pebble (Fleming, 1964).

In geomorphology therefore the decades of the 1950s and 1960s were characterized by an expanding range of subjects for attention, the import of quantitative techniques, and the application of those techniques. There was still an element of obscurity about the optimum purposes for which the techniques might be used and in the preface to one book (described by a reviewer as a recipe book, Brunsden, 1972, p. 260) It was stated that

> The increasingly accurate methods of measuring land form and geomorphological processes are providing a vast amount of quantitative data. This has to be analysed by numerical methods so that an orderly behaviour may be discerned from amongst the mass of accumulated data (Doornkamp and King 1971, p. v)

and in a reflective comment on the use of quantitative techniques (Thornes and Ferguson, 1981, p. 284) it was concluded that:

> The main application of quantitative methods has been and still is the description and analysis of field data on particular sites, events, or areas in a mainly inductive framework.

The characteristics of the import of numerical techniques into geomorphology are well exemplified in two singular developments. First, in what was probably the single most influential paper on the application of statistical methods in geomorphology, Chorley (1966, written in 1961) introduced the basis of measurement and sampling, and then the principal strategies available for the analysis of geomorphic data. In this paper it was acknowledged that statistical methods are only an adjunct and not a substitute for the initial qualitative stage of any investigation but that

> Statistical methods are tools which assist and test imagination. Some of these tools, however, like Galileo's telescope, may prove to be vehicles which enable human imagination and intellectual grasp to operate on higher planes than ever before' (Chorley, 1966, p. 377).

The second development was the attempt to progress towards spatial analysis in geomorphology and not only had the diffusion of quantitative techniques into geomorphology been regarded as slow by some researchers but also there had been a reluctance to embrace spatial techniques despite being heir to two distinct quantitative dynasties: one concerned with the application of quantitative techniques to geology pioneered particularly by Krumbein and the other the spatial analysis in human geography beginning in the late 1950s (Chorley, 1972). Reasons for delay in adopting spatial analysis techniques were adduced to be because many central geomorphic problems had not been traditionally expressed in spatial terms; many innovators of quantitative techniques had insufficient mathematical background to develop their use in geomorphology; there were few researchers available; in the United States and Sweden, where spatial analysis was most developed in geography, the links between human geography and geomorphology were traditionally rather weak, compared with those in Britain; work in automated cartography largely centred around technical problems of data storage and retrieval rather than use of data in model building; and spatial model building proceeds most rapidly where ideas of systems analysis have already been assimilated. In spatial analysis, viewed in relation to general, point system, network, continuous distributions, space partitioning and simulation aspects, the extension to other aspects of physical landscape such as Karst are very evident (Chorley, 1972).

It is of course very difficult to separate quantitative developments from progress in other fields and particularly from those relating to landscape processes (p. 87). However several papers with important general significance emerged in the early 1960s and although they came to have particular significance for studies of process and for systems approaches they were important influences upon, and to some extent were generated by, the inception of quantitative approaches. These papers included elucidation of the concept of entropy in landscape evolution (Leopold and Langbein, 1962) which dealt with the probability of various states but included examples from drainage networks and stream net simulation; and probability concepts in geomorphology (Scheidegger and Langbein, 1966) advocating a statistical or probabilistic viewpoint which could ascertain how theory would agree with empirical data. The influence of 'laws' elsewhere in science was apparent here, as many have been based upon the concept of minimization of effects (Williams, 1978). The basic underlying concept was that physical effects in the operations of nature, once having attained an equilibrium condition, change as little as possible afterwards by minimizing the effect of any subsequent disturbance and/or by minimizing energy expenditure. It is debatable whether the full import of these challenging developments was realized by geomorphological research in the succeeding two decades because even in the 1980s developments are occurring which may reflect and utilize such imaginative ideas (see p. 165).

A further important, but not well cited paper, referred to association and indeterminacy (Leopold and Langbein, 1963). Association was identified as the basis for geologic reasoning and was useful in indicating sequences of events in time; to extend from local to general; and to indicate processes although it was concluded that 'in geomorphologic systems the ability to measure may always exceed ability to forecast or explain' (Leopold and Langbein, 1963, p. 191). The principle of indeterminacy although long recognized in physics, was new to geomorphic thinking and referred to those situations in which the applicable physical laws could be satisfied by a large number of combinations of values of interdependent variables, so that the result of an individual case is indeterminate and although the range of uncertainty should decrease as more is learned about the factors involved, the uncertainty will never be removed. Leopold and Langbein (1963, p. 192) concluded:

> The measure of a research man is the kind of question he poses. . . . Geomorphology is an example of a field of enquiry rejuvenated not so much by new methods as by recognition of the great and interesting questions that confront the geologist.

Perhaps this is an admirable epitaph for any attempted summary of quantitative developments in geomorphology!

Biogeography

There were comparatively few quantitative contributions to biogeography by physical geographers before the 1970s partly because biogeographers were relatively few in number and also because they were directing energy towards historical aspects (p. 81) so that they tended to be more concerned with plants and animals in evolutionary time than with ecological time aspects (Simberloff, 1972) (p. 94). This paucity of contribution is reflected in the fact that there is no frontier described for biogeography in *Frontiers in Geographical Teaching* (Chorley and Haggett, 1965) and the chapter in *Models in Geography* was directed towards organisms and the ecosystem (Stoddart, 1967b) and eventually separated from the physical chapters when the book was reprinted in several paperbacks in 1969. It is also notable that pleas from some geographers (e.g. Edwards, 1964) that biogeography deserved a more significant place in geography curiously ignored the quantitative advances being made and the opportunities becoming increasingly available as a result of the greater attention accorded by ecologists to biological production by ecosystems. This was emphasized by the inception of the International Biological Programme in 1965. However, biogeography tended to emphasize spatial aspects and environmental relationships with insufficient early recognition of the potential application of quantitative methods in this regard. Thus the population biologists, MacArthur and Wilson (1967) thought that they detected a real difference between their approach to ecology and biogeography, which they categorized as concerned with:

> the limits and geometric structure of individual species populations and with the difference in biotas at various points on the earth's surface. The local, ecological distributions of species, together with such synecological features as the structure of the food web, are treated under biogeography only in so far as they relate to the broader aspects of distribution.

Quantitative advances being made in ecology embraced two groups of developments. First was theoretical deduction from mathematical models such as the theory of island biogeography, which involved a relationship between size of an island of area A and the number of species (S) of a given taxon in the form $S = CA^z$ where C is a proportionality constant dependent on the taxon and on the biogeographic region and where z ranges from 0.20 to 0.38. Second were developments in the statistical analysis of vegetation composition and a wide variety of numerical approaches, generated after the mid 1950s; of the order of 50 had probably been developed by 1970 (Moore, Fitzsimmons, Lambe and White, 1970). Some approaches endeavoured to subdivide vegetation into discrete components by association analysis which could then allow definition of vegetation units and the mapping of their area of occurrence. An alternative group of approaches depended upon the belief that vegetation varies continuously and is not susceptible to classification into discrete vegetation classes so that ordination procedures based upon gradient analysis were appropriate.

Although some of these approaches were later acknowledged in the researches of biogeographers, in 1978 five trends were isolated (Watts, 1978): investigations of first soil–vegetation–environment complexes; secondly of relationships between major vegetation types and particular animal species; thirdly analyses of distribution of individual species and of the influencing processes; fourthly of Quaternary community or ecosystem change; and fifthly of man–ecosystem–community relationships. Although the seeds of quantitative techniques were beginning to germinate in the work of biogeographers there was some delay before the productivity was very apparent!

Quantitative approaches required an even longer gestation period in the case of soil geography although this may have been because biogeography tended to subsume the study of soils until the 1970s when advantages were perceived for studying soil geography as a new and vigorous branch of the subject (Bridges, 1981). Jenny (1941) had provided an essentially quantitative perspective through the factors of soil formation, which though often cited was not developed by physical geographers who waited until the 1970s to develop the systems approach to the soil system. The quantitative approach to the soil profile was also somewhat delayed although there were attempts to rationalize data collection and mapping and to codify soil profile properties so that each soil profile could provide up to 200 items of information for computer analysis (Webster, Lessells and Hodgson, 1976). In soil geography the process of data acquisition especially by mapping, the dominance of researchers other than physical geographers, together with the data-rich character of the soil itself necessarily retarded the development of quantitative methods although techniques for analysis of soil distribution were reviewed by Courtney and Nortcliff (1977). They note that despite the great attention devoted to soil survey procedures it is perhaps alarming that most investigators have paid scant attention to spatial relationships. Although there is no single soil property which typifies the soil as a whole, classification, ordination, regression and analysis of variance methods can be used and merit further use in relation to studies of soil boundaries and soil variability (Courtney and Nortcliff, 1977).

Quantitative assessment

The quantitative impact in physical geography was registered in the decade of the 1960s but the subsequent repercussions of that impact must also be considered. A number of trends may be detected germinating in the 1950s, blossoming in the 1960s and creating seeds for publications and research developments in the 1970s and later. First was the evidence of the search for a focus beyond the historical emphasis which had prevailed until 1950. As this search coincided with the advent of enhanced computing power and the import of statistical methods into a range of sciences, it was inevitable that the methods involved in the search, and sometimes the focus itself, would be quantitative in nature. The extent of progress made varied in the several branches of physical geography and this variation has created an imbalance which has intensified and which has been the cause for a concern to be expressed by some writers. The degree of imbalance to some extent reflected interrelationships with other disciplines. Geomorphology attracted most physical geography research and although the disciplinary boundaries in the United States located most of the geomorphological advances in the geological fraternity, elsewhere geologists tended to ignore the Quaternary. Climatology and biogeography were much less evident in physical geography but had the benefit of a much greater rate of progress in related disciplines. Physical geography became less isolated from other sciences but concurrently there was a danger of its becoming divorced from human geography. Thus human geographers such as Gould (1973, p. 271) argued that most of physical geography is 'totally irrelevant to human spatial organization, except at the most obvious and naive level' and some physical geographers argued for a closer liaison of geomorphology with geology than with geography (e.g. Worsley, 1979). There is a danger that human geographers could make premature conclusions such as 'Physical geography, as most of us knew it, either survives as a second-rate earth science or has vanished completely' (Gould, 1973, p. 271), which may reflect a lack of knowledge of the complete range of work in progress (see p. 196).

Secondly, it was certainly the case that quantification in physical geography sometimes degenerated into the condition of new quantitative techniques in search of data. However, this was offset by many positive and exciting developments which stimulated thought about the purpose, nature and methods of the branches of physical geography. Paradoxically the advent of quantification may have intensified the separatist tendency despite endeavours to demonstrate the applicability of models, of networks and of numerical techniques throughout the several branches of physical geography. Models were generally classified as natural or conceptual, as analogue including physical hardware ones, and as theoretical which included statistical ones based upon probability and as mathematical dependent upon physical relationships. As models were assimilated into, and generated by, the branches of physical geography, the model-based approach was advocated (Chorley and Haggett, 1967) as the paradigm for geography to adopt. With the benefit of hindsight it is apparent that this view may have been unduly influenced by the single prevailing paradigm vision of Kuhn and that something more than models was required as the goal for physical geography. Two alternatives were developing: one amplifying earlier trends is reviewed in chapter 4 and another, which provided a greater break with tradition and was intricately intertwined with quantitative advances, is the subject for chapter 5.

4

Chronology continuing

Themes in existence by 1950 were continued in subsequent decades but they have developed so extensively and changed so dramatically that their origins may now appear to be many years away. Although the historical approach to physical geography had prompted an approach often styled as denudation chronology, the emphasis had been upon the sequence of denudation rather than upon chronology strictly connoted as the science of computing dates. In much work that was to develop within the compass of physical geography, the chronology assumed a greater significance, it required a greater knowledge of existing and recently developed dating techniques, and it thus involved close liaison with other disciplines particularly geology, biology and archaeology amongst the broadening spectrum of the earth and environmental sciences. Although it may seem that the developments were exclusively within geomorphology this was not the case, because environmental reconstruction was increasingly shown to be dependent upon knowledge of the ways in which the soils and sediments as well as the surface morphology related to past systems of vegetation and to patterns of climate.

When attempting to reconstruct the past evolution of physical landscape and to establish the chronology of its stages of evolution, we are confined to 'windows' of limited and varying opacity and size (Lewin, 1980). It is through the 'windows' that survive that we can glean the evidence necessary to permit environmental reconstruction and, necessarily, the amount of evidence surviving in a particular area will depend upon the age of the landscape and the subsequent changes to which the landscape has been subjected. The evidence that can be obtained through the available windows is of four major kinds. First is evidence relating to morphology of the environment, and a portion of a river terrace could be a fragment of evidence through the window of a much more extensive valley floor in the past. Secondly is evidence from sediments and materials, and the material comprising the river terrace could be used to infer something about the mode of deposition and the physical environment at the time. Thirdly is knowledge of the processes operating in the landscape and this may be gleaned from historical records or by analogy with situations elsewhere. Fourthly are the fragments of evidence that allow relative and absolute dating to be undertaken by an increasingly varied range of techniques.

In an early comment upon denudation chronology Day Kimball (1948) proposed that historical geology could be divided into two parts: stratigraphy, which deals with what is there and denudation chronology, which is concerned with what isn't! When concerned with denudation chronology it is necessary to utilize as much information as possible and to avoid conceptual approaches which are too constraining. When looking through windows for evidence it is often the case that the researcher only sees those types of evidence with a significance that has previously been established, and with an association that complies with existing models. Confronted by a section in a Quaternary deposit there are physical geographers who would devote all their time to the analysis of the sediment characteristics and not look at the space relationships of the feature in which the deposit occurs, and there are others who would deliberate about morphological evolution without closely investigating the sediment. It is obviously desirable that both approaches should be fully integrated and the *a priori* models (Figure 3.1C) should not constrain the information actually seen and obtained through the available windows! From 1950 to 1980 the tendency was for chronological studies to become increasingly sophisticated and more interdisciplinary so that this is a stricture on visualizing earlier work rather than the most recent. Although it could be argued that it is not really chronology continuing (as suggested by the chapter title) because there had been insufficient time specification before 1950, the sequence of this chapter proceeds from the basic foundations, to the alternative models, to sea level changes, Quaternary geography and hence to the prospect of environmental change.

Basic foundation

A feature of early denudation chronology was that it tended to concentrate upon particular areas and that the record deduced for those areas tended to exercise an unduly significant influence upon the way in which new areas were interpreted. Southeast England was the subject for a monograph of monumental significance (Wooldridge and Linton, 1939) and the sequence of landscape evolution deduced for this area came to have repercussions throughout subsequent research. Emphasis in the denudation chronology of southeast England centred upon drainage evolution and upon erosion surfaces which were later described as planation surfaces; relied upon interpretations based upon morphological evidence and information from fragmentary deposits; and produced a model which involved early Tertiary planation, mid Tertiary uplift, late Tertiary planation, early Quaternary planation by a Calabrian shoreline and subsequent valley development during Quaternary sea levels at successively lower altitudes. In the light of this model other areas were investigated including Wales (Brown, 1960), where the major upland surfaces were visualized as equivalent to the late Tertiary surface of southeastern England. The substantial amount of research in Britain in the 1950s and early 1960s generated a number of developments and debates. *Developments* included the realistic estimation of 11 million years as the likely time required for the production of an erosion surface and Linton (1957) had shown that the time scale used by some as a frame of reference for landscape evolution was too long. In his textbook Thornbury (1954) enunciated nine fundamental concepts of geomorphology of which one was 'Little of the earth's topography is older than Tertiary and most of it no older than Pleistocene.' A further development was to embrace the stratigraphic sequence in the conclusions about denudation chronology, an effective comparison was made of Britain and Appalachia and the two were embraced within a general model which acknowledged the Tertiary stratigraphy of both areas (Brown, 1961).

Denudation chronology also produced a number of *debates* which centred not only on the existence of surfaces of a particular number, which was not readily resolved by trend surface analysis, but also on the mode of development of planation surfaces and the significance of earth movement in influencing the denudation chronology. The principal debate about mode of origin initially centred upon the subaerial or marine hypothesis but it was later deduced, in the light of studies of contemporary shore processes, that marine planation could only be effective for less than one kilometre unless sea level was rising (Bradley, 1956; King, 1963). The model for southeast England has been refined and developed and one reason for its modification has been the time allocated to mid-Tertiary earth movements as it was appreciated that these 'outer ripples of the Alpine storm' may have continued for longer than was at first envisaged. Comparison of present rates of denudation and orogeny allowed Schumm (1963a) to deduce that the production of a planation surface could take between 10 and 110 million years and to show that present rates of orogeny are up to 10 times faster than rates of denudation. Such contrasts, although heavily biased towards certain areas of the earth's surface, underline the necessity of incorporating earth movement in the denudation chronology model and to appreciate that much of the progress in denudation chronology was achieved against the background of tectonically stable areas. This is much more pertinent as the tectonic background has been elucidated more completely (e.g. Vail, Mitchum, Shipley and Buffler, 1981; Evans and Hughes 1984). The approach of denudation chronology was assimilated in what Jennings (1973) characterized as 'the geomorphic band waggon parade' in which a series of fashions in geomorphology were listed as denudation chronology, climatic geomorphology, morphometry, general systems approach, process study and structural geomorphology. Structural geomorphology is still sometimes allied with the study of landscape evolution and Ollier (1981), in a view that would not be supported by all geomorphologists, has contended:

> To play a part in the problems of geology over the next few decades

> geomorphologists must forget their trivial catchments and see mega forests instead of trees. . . . Dynamic equilibrium, climatic geomorphology and process studies have all been shown to have limited application to geomorphology wherever geomorphic history is measured in hundreds of millions of years. If we reject cyclic ideas and even uniformitarianism, what have we left? The answer is evolutionary geomorphology.

It is true that geomorphologists have tended to neglect large-scale continental problems but this arises because they have not been prepared to contribute to tectonics and geophysics, which must surely be a prerequisite for an effective and accepted research contribution at the world level. It is evident that Ollier (1981), although making a plea for a more evolutionary geomorphology, does not cite the quantitatively expressed geophysical research and it seems unlikely that geomorphology can contribute in such areas unless it is willing to adopt the methods and language of other practitioners. Further pleas for macro-geomorphology have been made by Summerfield (1981) arguing for a more secure basis of geophysical, sedimentological and geochronometric, data and the potential for mega-geomorphology may remain (Gardner and Scoging, 1983). This is not to deny with Brown (1980) that geomorphologists have a part to play in elucidating the history of the shape of the earth, particularly in the light of the revolution that the theory of plate tectonics brought to the understanding of continental distribution. Brown (1980, p. 11) collects together seven principles which underlie historical geomorphological studies as:

1 Uniformitarianism
2 Evolution – the appearance of grasses, for example, in the Tertiary could have reduced erosion rates
3 Spatial variations in rates of operation of geomorphological processes
4 Base level – still crucial in historical geomorphology
5 Land surfaces are geologically young
6 Sediments provide the most important source of information concerning the erosional history of an area
7 Lithology and structure exercise a basic influence upon landform.

In many geomorphological histories there has been a propensity to ignore the most recent stages or time periods so that the Quaternary was often treated in less detail and the Holocene or last 10,000 years seldom referred to at all. To overcome this problem Brown (1980) advocated the use of a retrospective approach so that not only can the most recent stages be assimilated but also the knowledge of research on contemporary processes may be incorporated. In this way Brown (1979) devised five formative phases in the evolution of Britain which are necessarily unequal in length and almost plotted on a logarithmic time scale. Schumm (1968) also demonstrated how knowledge gained from contemporary denudation rates could give clues about the rates in geological time and suggested that with the appearance of grasses in the Cainozoic, the relations between climate, vegetation, erosion and runoff became much as today except for the subsequent influence of man.

Alternative models in historical geomorphology

Two broad groups of alternatives to, or extensions of, the denudation chronology model may be discerned, both dependent upon the influence of environment upon the researchers. First there were alternative models of historical geomorphology evolved in areas where remnants of Tertiary planation surfaces were dominant in the landscape and these are considered in this section; and secondly there were developments in areas where Quaternary deposits and landscapes promoted a very different approach and one which often centred upon sea levels and glacial landforms and their interpretation, as reviewed below (pp. 74 and 76).

A radical alternative to the peneplanation model that had been generated in humid temperate landscapes was pediplanation as conceived and developed by the geologist L.C. King. Developing from his early research in South Africa, Lester King constructed a new approach to geomorphology based upon a study of the world's plainlands (King, 1950) and in that approach claimed that continent-wide bevelled surfaces were produced by pediplanation. This provided a model for the evolution of African landscapes that depended upon parallel slope retreat and the coalescence of pediplains, and it stimulated canons of landscape evolution (King, 1953) and the extension of the model to many areas of the earth's surface in *The Morphology of the Earth* (King, 1962). In that universal approach, which provided a thought-provoking challenge to the established temperate models, King (1962, pp. 162–3, 643) introduced his model as:

> The classic account of the 'Normal Cycle of Erosion' as expounded by W.M. Davis has proved regrettably in error. With its emphasis on universal downwearing, it was a negative and obliterating conception resulting from cerebral analysis rather than from observation and has led to sterility in geomorphologic thought and retarded progress in the subject severely. The new pediplanation approach springs from the work of W. Penck, Kirk Bryan and Jessen. . . . (p. 643) the sequence of major cyclic denudation upon all the continents alternating with episodes of elevation and mountain building, together with the relations of both phases, through coastal plain and shelf deposits, with major events in the ocean basins, are not disturbed haphazardly through geologic time but are in broad temporal conformity one with another.

The model of pediplanation served to extend knowledge about the earth surface; to introduce greater cognizance of world landscapes because in his later papers King correlated surfaces from Australia, Africa and South America and in the *Morphology of the Earth* in 1962 embraced landscapes from the northern hemisphere as well; and to include earth movement in the form of cymatogenic arching as well as exogenetic processes as an integral part of the cycles of landscape development. Emphasis was still largely placed upon Tertiary rather than upon the most recent stages of Quaternary time. The King model of pediplanation provided a framework for an interpretation of the vegetation patterns of the savannas of the southern hemisphere and in a series of papers Cole (e.g. 1963) reviewed the climatic, edaphic and morphologic controls upon the types of savanna vegetation and concluded that the broad vegetation patterns related to the age of landsurfaces as expected from the King pediplanation model.

Also conceived in the tropics was a model of landscape development which was generated against the background of the seasonally humid tropics rather than the semi-arid, and which was related to weathering profiles and to igneous rocks and the experience of soil scientists together with the benefits of inputs from climatic geomorphology. This grew with the realization that deep chemical weathering would be the norm in many tropical areas and that many temperate areas included tor-like residuals which were a remnant from times when climatic conditions were warmer and wetter. In southwest England an early paper by Waters (1957) demonstrated the significance of differential weathering in oldlands and gave the basis for a new way of regarding the earlier stages of landscape evolution. It was records of deep chemical weathering which stimulated the development of new models and in particular the notion of a double surface of levelling (Budel, 1957). A lower basal weathering front marked the position at which chemical weathering was actively attacking sound unweathered rock, and on the land surface exogenous processes were eroding, transporting and depositing sediment across a landsurface composed of chemically weathered rock with occasional protrusions of unweathered erosional residuals. The relative behaviour of these two levels would reflect contemporary and past erosion systems and would vary from humid tropics to semi-arid and arid landscapes.

This significant model was adopted during the progress of subsequent research (see Thomas, 1978) and also had a significant relationship to climatic geomorphology. This type of geomorphology, which has subsequently attracted both great support and disenchantment, has been the subject of at least two groups of interpretations and it is perhaps the lack of equal familiarity with the achievements of both groups that may account for the differing viewpoints some of which are listed in Table 4.1. One view is that climatic geomorphology assumes that climate governs the character and distribution of landforms and although this view appears to be assumed by some of the critics of climatic geomorphology they do not cite many specific cases where the viewpoint is clearly advanced. However, this view arises because a klima-genetische geomorphologie was proposed by Büdel (1963) and became widely available in the English reading literature in 1965 (Holzner and Weaver, 1965). Büdel distinguished three generations of geomorphology namely dynamic which concerns the study of

Table 4.1: Views of climatic geomorphology

Date	Author
1979	C.D. OLLIER 'There is no doubt that climate affects landform development but it does not provide a good basis for a general theory of landscape evolution'
1980	I. DOUGLAS 'While climatic geomorphology proper might be seen as a valuable adjunct to applied pedology and applied ecology in understanding the operation of landscape systems at the present, climato-genetic geomorphology (Budel, 1963) which relies on the principle of actualism in historical studies is less readily tenable. . . . Geomorphological understanding has not been helped by the educational device of splitting geomorphology into two simple categories of climatic geomorphology and structural geomorphology which obscures the subtle interdependence between structural and climatic influences on landforms'
1968	D.R. STODDART 'While it may be possible to derive a satisfactory methodological basis for climatic geomorphology . . . one must conclude that, as at present formulated, much of climatic geomorphology is (a) not new, being implicit in Davis's own work; (b) not well-established, especially in terms of the climatic control of process; and (c) premature, in setting up generalized world-wide schemes, both of present climatic-morphologic regions, and of climatic-morphologic changes in the Pleistocene and earlier, before sound factual bases exist'
1973	E. DERBYSHIRE 'Traditional climatic geomorphology as represented by most of the papers in this volume has to a large extent glossed over this paucity of knowledge of fundamentals; it may be said to have proceeded, like Davis's work, to premature generalization on the basis of quite vague ideas on the underlying process relations. In the light of this, the way forward in climatic geomorphology, already recognized in studies of fluvial catchments, becomes evident. It will involve the geomorphologist increasingly in sophisticated instrumentation of both climatic and geomorphological parameters, the design of specific, long-term experiments and the use of multivariate methods. The requisite technology, plant and techniques are available and await the formulation of appropriate experimental designs.'
1976	K.W. BUTZER 'Climatic geomorphology attempts to cope with the excessive complexity of natural parameters by holding variables such as structure, lithology and man constant. In much the same way implicit or explicit models for describing the evolution of stream channels or drainage basins commonly are used to make simplifying assumptions that eliminate considerations of time, history and sometimes even progressive change.'
1980	J. BÜDEL 'The differences between various processes at work and the resulting differences among the landforms themselves are the object of climatic geomorphology.'

particular processes; climatic which considers the total complex of present processes in their climatic framework; and climatogenetic geomorphology which involves the analysis of the entire relief including features adjusted to the contemporary climate and also produced by former climates. This model was supported by five climato-morphogenetic zones (Büdel, 1963) which were later expanded to seven (Büdel, 1969) and then to eight (Büdel, 1977). Each zone was characterized by particular landscape-forming processes and by relief features and so could also be the basis for understanding past landscape development. Thus the extra-tropical zone of former pronounced valley formation was dominated by relict landscape features both glacial and periglacial. This model provided a way of visualizing the pattern of morphogenetic systems over the earth's surface during the Cainozoic and so lent itself to a clear way of integrating contemporary world zones with those of the past when ice sheets were non-existent or much less extensive.

The second view of climatic geomorphology emerged much more in relation to attempts to relate process to climate and to emphasize the interrelation between the morphological, pedological, vegetational and climatic characteristics of the earth's surface. This theme emerged them when Peltier (1950) identified nine different and possible morphogenetic regimes each of which should be distinguished by a characteristic assemblage of geomorphic processes. Peltier (p. 222) acknowledged earlier ideas when discussing the distinction of morphogenetic regions:

> The climatic boundaries of this graph are in part the same as those given by Penck (1910), Davis (1912) and Troll (1947) whose influence is acknowledged by the writer. Particularly are the glacial, selva and arid regimes . . . parallel to the glacial, humid and arid climates of both Davis and Troll. I should, however, say that many of the ideas here expressed are explicitly stated in the lectures of Professor Kirk Bryan whose emphasis on climatic morphology has led to the formulation here set forth.

He then proceeded to identify a periglacial cycle which went beyond a relationship between climate and process and attracted some of the criticisms that had previously been levelled at the Davisian 'normal' cycle. Somewhat independently, climatic geomorphology was developing in France under the influence of J. Tricart and A. Cailleux. A feature of the morphoclimatic zones recognized by the French school (Tricart, 1957) was the attempt not only to relate such morphoclimatic zones to climates and to processes but also to soils and to vegetation. Indeed the approach in some ways resembled that of the early Russian school of soil science and involved the recognition of zonal phenomena as the direct results of latitudinal climatic belts; of azonal phenomena arising from non-climatic control including endogenetic effects; extrazonal phenomena which occurred beyond their normal range of occurrence such as sand dunes on coasts; and polyzonal phenomena including those which operate in all regions of the globe subject to the same basic physical laws. This school of geomorphology produced an introduction to climatic geomorphology (Tricart and Cailleux, 1965, 1972) and a series of volumes dealing with specific groups of morphoclimatic zones. Although simple relationships based upon mean annual values of climatic parameters are prone to oversimplification, more recently this second approach has developed along more specific process lines (see p. 110) and has had much in common with energy budget climatology, particularly as pursued by the Russian school subsequent to the work of Budyko (1958).

A potential danger is that the trend away from the identification and analysis of planation surfaces could go too far so that insufficient attention is accorded to these remnants in the landscapes of areas dominated by formerly extensive landsurfaces. Particularly in parts of the world such as Australia the planation surfaces contribute a major part of the geomorphological story and Ollier (1979, p. 534) has concluded that process studies are of limited value; the present is not invariably the key to the past, much of Australia is not in equilibrium with the present conditions of base level, tectonics or climate, and cyclic theories do not fit the Australian scene very well. Landform inheritance (e.g. Pain, 1978) is therefore a concept emphasized by some researchers in environments of this kind, and Ollier (1979, 1981) proposes evolutionary geomorphology not as a cyclic approach with a sequence of stages but suggests that the earth's landscapes as a whole are evolving through time and this is analogous to the concept of an evolving earth as used in some geology books (e.g. Windley, 1977). Evolutionary geomorphology has been used with a very different connotation by Thornes (1983b) and this is introduced in chapter 8 (p. 182). In collecting together a Benchmark series book Adams (1975, p. 449) suggested areas of general agreement concerning planation surfaces:

1 Planation surfaces do exist and are not merely steady-state topographic products of random dissection.
2 the most distinctive unit surface is the pediment;
3 in attempting to estimate the emergence of a land mass from present altitudes of planation surfaces, their origins, base levels and most probably regional slopes must be considered;
4 most planation surfaces are or have been graded wth respect to some base elevation, usually sea level, but often to an endorheic basin or to a climatic control above sea level;
5 the most important time aspect of a planation surface is from the latest possible time of initiation of the cycle that produced it to the earliest possible time that it ceased being shaped (i.e. its terminal date) because of either burial or uplift;
6 a planation surface need not be presently under the same climatic regime as when it was initially developed;
7 planation surfaces should be considered in the general framework of plate tectonics, sea floor spreading, and the resulting change of continental areas with respect to the position of climatic zones.

These seven suggestions invite comparison with the 50 canons of landscape evolution that had been suggested by L.C. King in 1953 (King, 1953, pp. 747–50).

Sea level changes

A second group of alternative approaches was inspired in areas where Quaternary deposits and landscapes required a very different approach and this may be envisaged as of four basic kinds associated with sea levels, glacial sequences, periglacial landscapes and arid landscapes.

Sea level change inspired studies because many areas possessed evidence of stages of Quaternary erosion marked by river terraces and adjacent or other areas presented former raised shorelines around coastal margins together with evidence from buried valleys and remnants of former sea levels which were later submerged. Studies of this kind were influenced first by studies and methods of denudation chronology, of which they naturally formed the later stages, and secondly by developments in the study of the Pleistocene where the work of Zeuner (1945, 1958) and the work of geologists (e.g. Wright, 1937; Flint, 1947; Charlesworth, 1957) had been particularly influential.

In the analysis of shorelines and chronological stages many studies first followed Baulig (1935) and other workers, including Deperet working on Mediterranean shorelines, and then were influenced by the masterly synthesis achieved by Fairbridge 1961 and by Zeuner's (1959) consolidated view of sea level change in the Quaternary. In the latter view a classic sequence of Pleistocene sea level (Fig. 4.1) fluctuations included transgressions associated with interglacials and regressions accompanying the glacial phases and these fluctuations were superimposed upon a gradual sea level decline during the course of the Pleistocene. This explanation was primarily glacio-eustatic in nature because it depended upon sea level fluctuation in relation to the amount of water stored in the ice-caps during glacial and interglacial phases. Other world-wide eustatic causes of sea level change include sedimentary infilling of ocean basins which could give a sea level rise of 4 mm/100 years equivalent to 40 m in a million years (Higgins, 1965); orogenic eustasy whereby orogenic uplift creates ocean basins of different size; geoidal eustasy whereby the ocean surface reflects the variations in the geoid surface due to the earth's irregular distribution of mass which can give a difference between lows and highs of as much as 180 m. In addition, a number of local factors can be responsible for sea level change (Goudie, 1983) and these include glacioisostasy whereby the earth's crust responds to the development or removal of large ice sheets, and hydroisostasy when a similar response occurs as a result of large bodies of sea water or lake water from continental shelves and lake basins; orogenic and epeirogenic activity; compaction of sediments; and the increased gravitational attraction associated with large Pleistocene ice-sheets. The initial model proposed, based upon data from the Mediterranean, has been superseded with the advent of information from a greater range of sites elsewhere in the world, with emphasis upon correlation other than on simple altitudinal grounds and with more complete methods of absolute

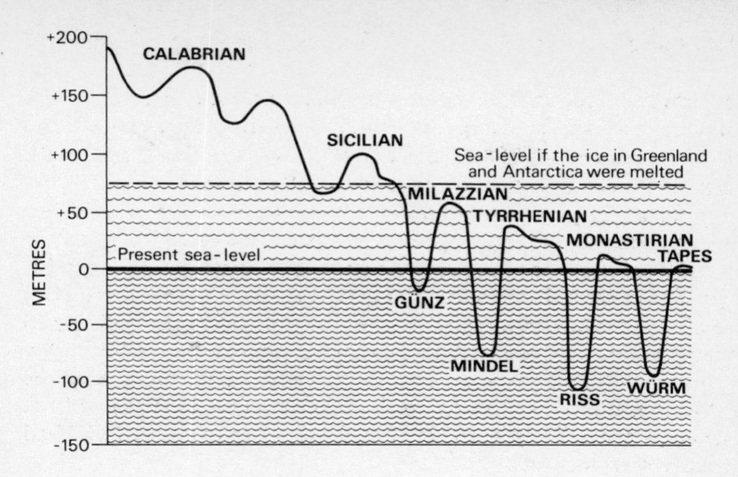

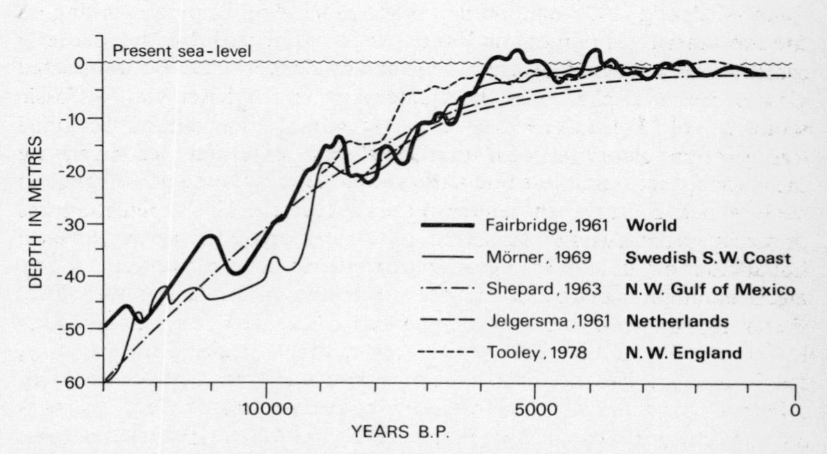

Figure 4.1: Views of Quaternary change (after Goudie, 1977) The classic sequence of Pleistocene sea levels is shown above and sea level curves compiled by various authors for the Holocene are indicated below.

and relative dating. Of the latter the advent of oxygen isotope dating of deep-sea sediment cores has probably been most significant and also very important in relation to reconstructing climatic change and glacial chronology. In addition, the use of Uranium-series dating of uplifted coral coasts (Chappell. 1974) and of amino-acid dating, especially as applied to coasts of the USA (Wehmiller, 1982), has advanced knowledge and has allowed separation of sea level and tectonic components on rapidly uplifting coastlines. However, it is still not possible to be certain about world sea levels before the last quarter of a million years. Perhaps it is not unexpected that simple correlation cannot easily be made between different parts of the world in view of the advent of the theory of plate tectonics because of the amount of movement associated with the displacement of plates. Particular attention has been devoted to the Late Quaternary sea levels and especially to those of the Holocene. In consolidating data from several areas it has been proposed that a rapid rise of sea level occurred in the early Holocene but that in the last 6000 years the rate has been far less, although it may have diminished progressively, it may have risen to *c.* 3600 BP and then remained constant, or it may have oscillated to positions above and below present levels (Fig. 4.1). In an extensive review arising from IGCP 61 Kidson (1982) has concluded that the search for a universal eustatic curve should be regarded as over; that regional differences in changes in the geoid mean that eustatic sea level curves can have only regional validity; and that no part of the earth's crust can be regarded as wholly stable. Work on sea level changes has tended to concentrate upon these most recent stages (e.g. Tooley, 1978) with a considerable effort devoted to isostasy (Andrews, 1970a). In Scotland detailed investigation of raised shorelines as summarized by Sissons (1976) has revealed how a complex pattern of late Devensian and Flandrian raised shorelines have been differentially uplifted and are slightly diachronous because the parts nearer to the centre of uplift were lifted clear of the sea earlier than the peripheral remnants (Smith, Morrison, Jones and Cullingford, 1980). This view as a result of very detailed levelling has refined the earlier interpretations of the distribution of post-glacial shorelines and has provided vital information to assist in the interpretation of late Quaternary history, often relying upon an effective blend of techniques from geomorphology and biogeography.

Quaternary geography

In areas dominated by Quaternary deposits emphasis in research in the late 1950s and subsequently was naturally directed towards the interpretation of the stages and nature of Quaternary landscape change, and this involved the inception of new conceptual models of glacial landform development well exemplified by the way in which J.B. Sissons of the University of Edinburgh developed a theme in many research papers of the interpretation of patterns of deglaciation involving stagnant as well as active ice (Sissons, 1976). However, three major features may be discerned as characterizing research by physical geographers on the Quaternary: increasing involvement in chronology, the impact of new techniques, and the progress towards an interdisciplinary focus.

When influential textbooks were produced by Quaternary geologists (e.g. Flint, 1947; Charlesworth, 1957) the disparities between areas such as northern Europe, central Europe, and North America were perhaps most apparent. A classical model was produced by Penck and Bruckner in 1909 and this involved four main glaciations although it has subsequently been appreciated that the sediments on which this four-fold sequence was based represented only a small proportion of the time span, and unconformities between each successive terrace probably conceal events lost to the record regionally or locally (Bowen, 1978). The four-fold sequence, and the four names derived from the names of Alpine rivers, continue to echo in research around the world although this interpretation has now been superseded. Additional models had been proposed for Northern Europe where a classic sequence of glaciations and interglacials had been proposed based on the investigation of sequences on the North European Plain and in Scandinavia; in the British Isles; and in central North America (Bowen, 1978). Although physical geographers had not contributed to their development, these model sequences were increasingly evident in research writings as knowledge of the sequence in one area and correlation between areas were necessary to proceed towards explanation of local and then of regional patterns. Whereas influential books like Flint's *Glacial and Pleistocene Geology* adopted a structure in which landforms and landscape patterns comprised the early part of the book and then were succeeded by chapters on stratigraphy and chronology in different areas and this was echoed in the structure of research on specific areas; by the late 1970s books by physical geographers had adopted a structure in which chronology and the classic models appeared at the beginning.

Greater interest in, and involvement with, chronology and environmental reconstruction became possible because of the greater pace of development of new techniques to complement established ones which could be used for dating and/or the provision of environmental information. Such established techniques included the analysis of varves in lake deposits indicating an annual or seasonal rhythm, used since the pioneer work of de Geer in Sweden in 1912; analysis of tree rings when dendrochronology has been linked very convincingly to climate; relative dating of landforms; and palynology and other types of micro-fossil analysis including non-marine mollusca. Some of these techniques have been extended and especially at the University of Birmingham G.R. Coope (Coope *et al.*, 1971) has made great progress as a geologist who used the wing cases of fossil beetles (Coleoptera) to indicate climatic characteristics of palaeoenvironments because coleoptera are very sensitive to changes in wetness and dryness. Palynology has been adopted by many biogeographers and pollen analysis was the basis for the reconstruction of the detailed vegetation history of many specific areas although more recently emphasis has changed from individual sites to the reconstruction of patterns of change in Britain (e.g. Barber 1976) and in the tropics (e.g. Flenley, 1979).

To complement these existing techniques new ones have developed which derive particularly from isotopic dating techniques and from analysis of deep-sea ocean cores. Isotopic methods provide geochronometric methods of dating to support stratigraphic investigations which are radiometric to complement the use of varves, dendrochronology or palaeomagnetic evidence that can give chronology relative to a floating scale. Radio-carbon or C14 dating was first applied in 1949 and together with other evidence now provides the chronology for the last 50,000 years and may be extended to 75,000 years. As growing organisms incorporate radio-carbon and after death of the organism the trapped radio-carbon begins to decay at a known rate with half lost after 5730 years, it is possible to indicate when death occurred. Such radio-carbon dating has been used on wood, charcoal, peat, organic mud and calcium carbonate in molluscs, foraminifera and bones. Uranium nuclides ^{238}U and ^{235}U decay to stable lead and this decay is the basis for a method applied to molluscs, coral and deep sea sediments which usefully complements ^{14}C dating because it can be used for materials up to 350,000 years in age. Potassium-argon dating can be applied to volcanic rocks such as lavas and tuffs and although difficult to measure the decay, it attains its maximum usefulness in the Middle and Early Pleistocene and can be used for a range greater than 20,000 years. The range of techniques has been reviewed by Bowen (1978) and by Goudie (1981a) and an indicative summary is included in Table 4.2.

Table 4.2: Some developments of techniques for dating and environmental reconstruction (*Based upon Bowen, 1978, Worsley, 1981; and Cullingford, Davidson and Lewin, 1980*).

Technique	Methods	Range
Radiocarbon dating	Measurement of radioactivity in a fossil sample compared with modern standard and knowing rate of decay (half-life).	Wood, charcoal, peat, organic mud, calcium carbonate in molluscs, foraminifera and bones, soil organic matter
Uranium series disequilibrium methods	Two long-lived Uranium nuclides ^{238}U and ^{235}U decay through a series of short-lived products to stable lead.	Fossil corals
Potassium-Argon and Argon-40/ Argon-39	Small amounts of ^{40}K in rocks decay to ^{40}Ar and the ratio of the two can give a measure of age of rock.	Volcanic rocks including lavas and tuffs
Fission track dating	Traces of ^{238}U in minerals and glasses of volcanic rocks decay by spontaneous fission at a known rate.	Volcanic rocks and provides checks on other methods.
Magnetostratigraphy	Analyses of natural remanent magnetism (NRM) in rocks which reflects orientation and intensity of earth's magnetic field at time of rock formation.	Volcanic and some sedimentary rocks
Amino acid diagenesis	After death of organisms amino	Bone samples in

Table 4.2—(*contd.*)

	acids in protein are released and the ratio of amino acids can give an indication of relative age.	uniform temperature conditions
Glacial varves	Analysis of accumulation of annual layers accreted in still water.	Lakes in proglacial areas
Dendrochronology	Annual growth rings give indication of number of years and wet summer may produce thicker rings	To indicate climatic variations and age of surfaces such as slopes
Stable isotope fractionation	Small but measurable differences in isotope composition of any two chemical species of the same element. Thus differences in $^{18}O/^{16}O$ ratio may be expected between water and calcite precipitating from it.	Deep sea sediments, Continental freshwater calcareous deposits, water in hydrological cycle
Thermoluminescence	Some materials can acquire thermoluminescence from irradiation by naturally occurring radio isotopes and length of exposure can be deduced.	Pottery at archaeological sites, some Pleistocene sediments including loess
Obsidian hydration	Rhyolitic volcanic glass (obsidian) when exposed to the atmosphere absorbs water to form a hydrated rind that increases in thickness in relation to time and/or temperature.	Archaeological samples of obsidian artifacts
Lichenometry	Measurements of lichen thalli on surfaces of known dates can be basis for growth curve which can then be basis for indicating age of other surfaces.	Depositional surfaces, and features such as moraines and erosional surfaces such as rock fall scars
Caesium-137	Isotope generated artificially by fission reaction and thermo-nuclear weapon tests since 1954 have given cascading systems of radioactive products in atmosphere, surface, soil and vegetation.	Recent sediment changes

Deep-sea cores potentially provide a complete stratigraphical record whereas the continental record is necessarily less perfect. With the development of appropriate piston corers in the late 1940s it was possible to collect columns of sediment 10–30 m in length which contained material that could be dated by radiometric and other means and could also provide environmental information, for example by analysing the frequency of sensitive foraminifera. Abundant information has been yielded by cores and one extracted from the Pacific near the equator from a water depth of 3120 m indicated a sequence of 900,000 years including eight completed glacial cycles and nine terminations which are rapid deglaciations. The new hydraulic piston corers (Shackleton and Hall, 1983) are now providing better samples as the basis for more detailed analysis to allow a greater degree of isotopic structure and this should enable the climatic variability of the early Pleistocene to be analysed in detail comparable with that already achieved for the late Pleistocene.

The frequency of glaciations indicated by this record has now been confirmed by the loess record in central Europe (Kukla, 1975) the USSR and China (Goudie, 1983). Much of the progress achieved in studies of Quaternary environmental change has involved increasing amounts of interdisciplinary cooperation. This has been manifested in a number of ways including the birth of interdisciplinary journals such as *Quaternary Research* (1970—) and *Boreas* (1972—) and *Quaternary Science Reviews* (1982—); of national organizations for Quaternary Scientists including AMQUA in America, DEUQUA in Germany, NORDQUA in Scandinavia and the QRA (Quaternary Research Association) in Britain; of the growth of the International Union for Quaternary Research (INQUA) with its meetings every 4 years and many commissions which coordinate research in many fields; and of international cooperation under the auspices of the International Geological Correlation Programme (IGCP) which includes Quaternary investigations within its endeavours. In addition the closer links between branches of physical geography such as geomorphology and biogeography have necessitated links with biologists and geologists and then with palaeoclimatologists because of the importance of studying climatic change. Perhaps the clearest indication of the increasingly multidisciplinary nature of Quaternary investigations is that a book written by a geographer, D.Q. Bowen, and published in 1978 is entitled *Quaternary Geology* and is subtitled *A Stratigraphic Framework for Multidisciplinary Work*. It is also notable that Bowen has been in the vanguard of interdisciplinary and international cooperation which has establish the utility of amino acid dating, and he has been instrumental in establishing a laboratory for this purpose in the Department of Geography at the University College of Wales, Aberystwyth.

The benefits of more extensive methods for dating and environmental reconstruction were also reflected in the investigation of Quaternary morphogenesis in areas beyond ice sheets. Although emphasis was initially upon recognition of the variety of landscape features, sediments and structures that could be developed under periglacial conditions, the potential subsequently arose of developing a greater knowledge of phases of periglacial landscape development, and in Poland and other countries in Europe this emphasis was clearly evident in research in the 1960s and much of the research was reflected in *Periglacial Geomorphology (Embleton and King, 1975)* which was one of two books to derive from the earlier *Glacial and Periglacial Geomorphology* (Embleton and King, 1968).

Earlier developments by geographers which proceeded somewhat independently were located in arid, semi-arid and subtropical areas. These were additional to interpretations of planation surfaces and were particularly concerned with the alternation of pluvial and arid phases in the Quaternary and their possible relationship to those of temperate latitudes; with the sequence of valley development particularly as revealed by the chronology of alluvial deposits; and with the significance of human activity which in some of the areas studied extended over many centuries. Such studies are exemplified by the work of Professor K.W. Butzer on areas in the Middle East and in the Mediterranean where he reconstructed stages of landscape change as related to Quaternary stratigraphy and indicative of changing Quaternary climates. These studies depended upon detailed analysis of fluvial, lake, aeolian and cave sediments together with inputs from related disciplines particularly to furnish techniques and results from palaeobotany, palynology, palaeoclimatology and archaeology. The latter was necessary because man-land relationships provided necessary insight into the use of environments throughout the middle and late Quaternary (see Butzer, 1964). A stimulating approach to *The Mediterranean Valleys* provided by Vita-Finzi (1969) also relied upon the close association between human activity and valley evolution as exemplified in the sediments in valley floors and revealed a fascinating sequence. In hot desert areas investigations of fossil dunes in India, Africa and Australia together with changes in the extent of pluvial lakes in Africa, the Middle East, and North America (Goudie, 1983) have provided further insight into the chronology of Quaternary environments.

A historical emphasis in biogeography prevailed during the 1960s in particular and is collectively described as Quaternary ecology by Simmons (1980). Using palynological and chronometric dating techniques a great number of cores have been investigated and the flora and fauna of the terminal Pleistocene and of the transition during the early Flandrian in western Europe from tundra to closed deciduous woodland, have been investigated and the major phases distinguished. Progress in this direction has now been achieved for a diversity of world areas and requires ancillary information related to climatic change and, in turn, when sufficient information becomes available could contribute to the reconstruction of world patterns. An excellent example of international cooperation is provided by IGCP Project 158B devoted to the study of lakes and mires in the temperate zone (Berglund, 1983) and to the collation of detailed information that facilitates the development of generalizations about trends in vegetation change and hence in climate during the last 15,000 years. This international project is organized from a Department of Quaternary Geology but involves inputs from physical geographers as well as from biologists, geologists and archaeologists. Climatology as viewed by a geographer necessarily involves consideration of climatic change (Lockwood, 1979a) and of the energy balance, general circulation, and statistical dynamical models (Lockwood, 1983a) that afford the main types of model used in studies of climatic change.

Changes of climate have also been investigated during the historic time scale when a range of historical techniques have been utilized, including diaries and records together with sedimentary evidence, and information from faunal remains, archaeology, tree rings and ocean deposits, and G. Manley, a geographer at London University prior to his appointment as the first Professor of Environmental Sciences at Lancaster University, was one who researched many of the obscure details of climatic fluctuations in Western Europe (e.g. Manley, 1952). Biogeographers investigating landscape change on the most recent time scale have also resorted to documentary and field evidence because palynology may not be sufficiently reliable at this scale for the studies which have been completed including those of the outer Leeward Islands (Harris, 1965), the southwest of the United States and the vegetation of Barbados 1627–1800 (Watts, 1966). In the development of studies of change particularly by biogeographers, increasing interdisciplinary cooperation and a greatly expanded range of techniques are very evident. In some cases geographers are adapting techniques, as in the use of soil micro-organisms as an indicator of the relationship between land use change and soil characteristics (Maltby, 1975) and developing techniques with a great range of applications based upon magnetic susceptibility (e.g. Oldfield, 1983a) and dating using ^{210}Pb which is of great accuracy and value for recent sediments (Appleby and Oldfield, 1978), and particularly enhanced the study of lakes and their drainage basins (Oldfield, 1977) and these developments are referred to also in chapter 6 (p. 124).

Quaternary geography and the prospect for environmental change

Studies of the Quaternary by physical geographers since the 1950s have led generally in one of two directions: either towards a more process or model-based investigation or towards a greater emphasis upon chronology. The former will reappear in chapter 5 in relation to investigations of glaciers, for example, and the latter is the reason for interdisciplinary cooperation that has been stressed in the preceding pages and is further developed in chapter 8. Such cooperation can present problems for teaching, training and research because the necessary techniques and methods are the province of more than one discipline. One solution has been to create university departments of integrated environmental science or of earth sciences. In some of these the emphasis is upon environmental change in the Quaternary whereas in others they have accommodated investigations of contemporary environmental processes as well. In the context of geomorphology Clayton (1971), himself a geographer progressed to environmental sciences at the University of East Anglia, has argued that it is difficult to achieve a balanced development across the two subjects of geography and geology that it depends upon, and Worsley (1979) particularly concerned for the diminishing resources available and the imminent policy of selective resource concentration argued:

> The inference to be drawn is obvious; regrouping is necessary and the longer the fateful day of decision is delayed, the greater the likelihood that the established

geoscience groups will capture those limited resources. It is hoped that the proverbial geomorphological cuckoo in its geographical nest will stretch its wings and fly, rather than become a grounded ostrich.

To continue the analogy however, one danger of a cuckoo in a nest is that before flying it may permanently displace other promising chicks which could be particularly concerned with processes (chapter 5) and man (6) and such themes are not readily accommodated in an earth science, geoscience or environmental science structure. Perhaps the answer lies in diversity and in that the very existence of different organizational structures in different countries is a very positive attribute and that when diverse arrangements exist within a single country this also may be highly desirable. Some physical geographers have perceived the need for adjustment of fundamental outlooks and Bowen (1979, p. 167) suggested:

> Outmoded models are still conventional wisdom and in geomorphology there are signs that, rather than face the new realities, there is a retreat into the more rarefield atmosphere of conceptual model-building and complex correlation structure diagrams. Or alternatively the new situation is completely ignored and the time–space mesh contracted to such a degree that *landform* is almost taken as read and interest is, instead, directed at hydrological matters relating to the transmission of water and sediment across land-surfaces which look increasingly like the isotropic ones of human geographers.

Similary Andrews and Miller (1980) have argued that the importance of chronology is frequently overlooked and/or understressed in many geomorphological studies and that

> The dominance of studies on modern processes in the field of geomorphological methods needs to be balanced by the longer perspective that falls traditionally within the area of Quaternary geology and geomorphology. . . . Chronology is one of the cornerstones of geomorphology, and researchers are urged to balance their studies of present geomorphic processes by an appreciation of rates of work over timescales of 10^4 – 10^6 years.

Such restraining views are typical of many that are fully justified by the emphasis upon process studies that characterized the 1960s and 1970s and led to smaller investment of resources in historical studies and to the fear that process studies will not profitably link back with chronological ones – in fact there are many indications that such links are happening again including the way in which physical geographers who have been mainly concerned with processes and modelling are moving towards evolution. One leading proponent, J.B. Thornes, in contributing to an interdisciplinary symposium in Geography, Archaeology and Environment, affirmed that palaeoenvironmental reconstructions are crucial if the understanding of the ecology of past societies is to be substantially improved (Thornes, 1983a) and elsewhere (Thornes, 1983b) has advocated evolutionary geomorphology and has anticipated that it will produce major new insights into historical problems such as river terrace formation and his proposal is considered more fully in chapter 8 (p. 182). G.H. Dury. a geomorphologist who has made a substantial contribution which includes a considerable number of process-type contributions as well as some concerned more with evolutionary geomorphology, concluded after a review of three recent statements on the issue (Dury, 1972, p. 201):

> My own response to the question of where geomorphology belongs must be, that the question is badly structured. Formally it belongs where its practitioners are attached for payroll purposes. Functionally, it belongs on the surface/subsurface interface. There was a time when the soil nexus held the promise of unifying physical and human geography. Now, the union if it is to come, might conceivably come through common techniques of analysis. Alternatively it might well come through environmental research, including work in environmental geomorphology. It seems highly unlikely to result from the subdivision of territorial claims.

Strictures such as those by Bowen and by Andrews and Miller are very necessary if physical geography is to avoid losing sight of the historical dimension and the links with process and systems studies (see chapter 8, p. 161). However, it may be that the existence of diverse approaches is inevitable and Butzer (1973) concluded that pluralism was inevitable and possibly desirable. There is still disagreement as to the extent to which the study of contemporary present-day phenomena can be a key towards understanding similar phenomena in the remote past. This is referred to as the principle of *actualism* by Douglas (1980) citing Russian geologists Gorshkov and Yakushova and, whereas it is rejected by Douglas and also by Ollier (1979, 1981), it is possible to believe that although the past development of an area may not be explicable in terms of the processes at present operating in that area, nevertheless study of present processes may collectively help to illuminate the past. In this case pluralism as visualized by Butzer is required. Indeed Butzer is a geographer who has made great contributions in defining a field of *Pleistocene Geography* which was the term used as a subtitle for his book on *Environment and Archaeology* first published in 1964. Whereas Pleistocene geology was seen to be primarily concerned with stratigraphy and chronology, Butzer saw the need for a more comprehensive study of past environments and envisaged Pleistocene Geography as concerned with the natural environment and focused on the same themes of man and nature that are the concern of historical and contemporary geographies. In addition to collecting the techniques available Butzer (1964) applied the zonal concept to provide a background of world physical environments against which he reviewed Pleistocene environments of the Old World and man–land relationships in prehistory. Subsequently a development of the theme has appeared in *Archaeology as Human Geography* (Butzer, 1982), in *Geoarchaeology* (Davidson and Shackley, 1976) and in a book which endeavours to develop the theme of the nature of the environment during each of the phases of British prehistory (Simmons and Tooley, 1981). A slightly different but very stimulating approach also originating from earlier experience of chronological change is a review of *Recent Earth History* (Vita-Finzi, 1973) which looks at the record particularly of the last 20,000 years and at methods of dating and clearly argues that:

> The immediate past does, of course, have the attraction of serving as a bridge between present day processes and their fossilized counterparts. Several themes in physical geology suffer from falling between the stools of historical geology and geomorphology. To judge from the literature, the distinction drawn between palaeontology and genetics or archaeology and anthropology, is sometimes equally deleterious even if largely administrative (Vita-Finzi, p. 111).

Quaternary or Pleistocene geography, environmental change (Goudie, 1983) geoarchaeology, and alluvial chronology are all examples of the new focus for chronological investigations. This new focus has required the replacement of conservatism, which in geomorphology has ignored subjects of topical concern like accelerated soil erosion, and the persistent misconception that denudation chronology can be equated with Davisian cycles 'dated' by landform geometries (Butzer, 1980). Butzer outlines a revolution in historical geomorphology that was made possible by the successful application of sedimentological techniques, by advances in isotopic dating, and by collaborative geo-archaeological efforts. He reviews the problems of internal dating, of intraregional correlation, of regional delimitation and of causation against the background of detailed results from a variety of areas proceeding towards a challenging opportunity to examine the spatial and temporal dimensions of human interference in complex ecosystems.

The need for a radical change of outlook has also been stressed by Bowen (1979) who sees progress in this direction being made by a mixture of basic and applied work by both multidisciplinary and interdisciplinary endeavour diminishing what Butzer (1975a) saw as particularism – the tendency for individual specialists or groups to regard their own field as the cornerstone for others. He envisages that a change in outlook may derive from appreciation of the complex event sequences that new techniques have now exposed in the Quaternary; from appraisal of the classical models of change to accommodate the realization that extreme rapidity of change now has to be considered when evaluating chronological biotal and geomorphological processes; from adjustment of geomorphology to new knowledge of Quaternary change such as rate of ice sheet growth and decay; and similar adjustment of biogeography and of palaeoclimatology. One of the most advanced examples of development has been CLIMAP (Climate, Long Range Investigation, Mapping and Prediction, Hays and Moore, 1973) which has made a reconstruction of global sea surface, ice extent, ice elevation and continental albedo for the northern hemisphere summer at 18,000 years BP and this has provided the boundary conditions for global general circulation models to simulate the climate when the northern ice sheets were at their maximum extent (see Lockwood, 1983a). Thus the inception of a new focus for the study of chronology has been promoted by such stimulating developments and Bowen (1979) points out that the CLIMAP group have discovered significant evidence which supports the idea that the immediate future is one of adverse orbital geometry and general cooling and hence that 'the prediction of the future must rest on the past in the present' (Bowen, 1979, p. 181).

5

Processes prevailing

Although study of the past can be useful for extrapolation into the future it is also imperative that investigation of contemporary processes is undertaken to assist understanding of the way in which environment operates now, operated in the past, and may operate in the future. Studies of processes were undertaken to remedy what many physical geographers perceived to be a deficiency and in doing so it was necessary to overcome considerable resistance such as that in geomorphology where Wooldridge (1958, p. 31) avowed:

> I regard it as quite fundamental that geomorphology is primarily concerned with the interpretation of forms, not the study of processes.

It may help to see how the need for process studies was perceived to exist and then to proceed to indicate in which branches studies were undertaken using what methods as a basis for assessing how our knowledge has increased in spatial terms and in time with reference to catastrophic views.

Perceiving the need for process

A recurrent criticism of historical approaches in physical geography up to the 1950s was that they had insufficient knowledge of environmental processes to fall back upon which could enhance the understanding of landscape. More specifically, in geomorphology it had often been pointed out that although the Davisian trilogy embraced structure, process and stage or time the emphasis had been very clearly upon stage with very little upon process. In an influential paper on the dynamic basis of geomorphology A.N. Strahler (1952, p. 924) of Columbia University characterized the situation:

> The weakness in understanding of geomorphic processes (and hence also a weakness in the understanding of the origin of landforms) has not been confined to the American continent. The geomorphologists of France and England, closely attached to the schools and departments of Geography, have also tended to give much attention to descriptive, deductive studies of landform development and to regional geomorphological treatments. . . . If geomorphology is to achieve full stature as a branch of geology operating upon the frontier of research into fundamental principles and laws of earth science, it must turn to the

physical and engineering sciences and mathematics for the vitality it now lacks.

Strahler continued (1952, p. 937) to suggest a programme for further research in geomorphology and this required five steps which were (1) a study of geomorphic processes and landforms as various kinds of responses to gravitational and molecular stresses acting on materials; (2) quantitative determination of landform characteristics and causative factors; (3) formulation of empirical equations by mathematical statistics; (4) building concepts of open dynamic systems and steady states for all geomorphic processes; and (5) deduction of general mathematical models to serve as quantitative natural laws. This programme was vast but necessary because Strahler suggested that geomorphology was already a half-century behind developments in chemistry, physics and the biological sciences.

This prescription for geomorphology was applicable to physical geography as a whole. It is important to make the distinction between what Chorley (1978) identified as functional and realistic approaches, because both were integral parts of the movement towards a greater focus upon processes. Functional studies were essentially positivist in nature and depended upon the notion that phenomena can be explained as instances of repeated and predictable regularities in which form and function can be assumed to be related, and indeed form and process figured in the titles of a number of books concerned with processes (e.g. Carson and Kirkby, 1972; Gregory and Walling, 1973). The realist approach, although developed as an extension of the functional positivist approach, attempted to probe beyond the relationships derived from observed regularities and to seek the mechanisms and underlying structures which are responsible for the operation of environmental processes. These functional and realist stages tended to develop in sequence and characterized each of the branches of physical geography albeit to varying degrees.

Processes were scrutinized to remedy deficiencies internal to physical geography. Therefore it was first necessary to gain information on the rates at which processes operate but then subsequently to establish relationships between process and controls. These could be utilized to estimate values for unmeasured situations or for unknown time periods despite the black box nature of this functional approach, and finally there were endeavours to proceed towards a more realist explanation. Within geography the replacement of an idiographic by a nomothetic approach produced an atmosphere in which process measurement was very germane. Furthermore, some branches of physical geography had proceeded as far as they could without an enhanced knowledge of processes. Thus the understanding of glacial landforms such as cirques required further knowledge of the processes of ice movement and glacial erosion and the interpretation of planation surfaces as having been produced by marine erosion required knowledge of the nature and rate of processes of coastal erosion. A further characteristic of earlier physical geography had been the tendency to ignore the Holocene and to concentrate instead upon earlier phases of landscape development. With attention turning to the Holocene it was desirable that more should be known, for example, about processes of ecological dynamics. Thus the historical approach to the development and distribution of savanna vegetation which had been developed by M.M. Cole had emphasized the relationships with landsurfaces of different age and character, but to demonstrate exactly how ecological processes were manifested it was necessary to focus upon the interrelationships of all factors in the savanna ecological system and therefore the dynamic ecology was a feature of the University of McGill Rupununi Savanna Research project (Hills, 1965, 1974) and of work by Eden (1964, 1974).

There were also indications that branches of physical geography were acknowledging the necessity for studies of processes in view of the way in which processes were featuring more prominently in related disciplines. This is reflected in the intentions of journals and in 1947 in the *Journal of Glaciology* the content of a leaflet issued by the British Glaciological Society was recalled in which

> Snow and ice from their mutuations and from the part they play in nature as precipitation, agents of erosion and modifiers of climate come within the range severally of physics, meteorology, geology, physical geography, oceanography and climatology. The behaviour of ice crystals in glaciers has close connection with crystallography and metallurgy.

Similarly in hydrology the *Journal of Hydrology* (1963–) arose because:

> The steadily rising interest in many countries in hydrology in all its branches makes this an opportune time in which to launch a new Journal concerned with the scientific aspects of the subject. There is too a growing appreciation of the need for an international approach to hydrological problems. . . .

In soil science emphasis had naturally proceeded towards soil processes including soil biology and microbiology (e.g. Russell, 1957) as described in the *New Naturalist Series* of books and interest in soil processes and soil dynamics also led to the inception of new ideas. Thus the factors of soil formation as propounded by Jenny (1941) was superseded by a more process-based vision of the soil profile as based upon additions, subtractions, translocations and transformation of constituents (Simonson, 1959). A further development was that the soil series adopted in the 1950s as soils with similar profiles, derived from similar materials under similar conditions of development as the basic unit for field mapping acquired a third dimension with the definition of the three-dimensional soil body or *pedon* (Johnson, 1963). The smallest volumes, usually between 1 and 10 m^2 on the surface, were the basis for mapping units, or *polypedon*, which contain more than one pedon. A further aspect of the climate external to physical geography was provided by the beginning of greater public awareness of the environment and the realization of the implication of the possibility of finite resources and that spaceship earth required concern for the wise use of its resources and reserves. Although this intellectual atmosphere, or aspect of social concern (Stoddart, 1981) was particularly germane to the study of the effects of human activity (chapter 6, p. 116) and to the application of environmental investigations (chapter 9, p. 186), it also created a situation in which acquisition of information on environmental processes and environmental change was certainly favoured and sometimes positively encouraged. Thus in the UK the Natural Environment Research Council had initiated a series of working groups to advise on the needs for future research. These included a working group in Climatology which suggested (NERC, 1976):

> The study and understanding of past climatic changes poses many problems of considerable scientific interest, but of great intrinsic difficulty. The need to assess the likelihood of possible future changes in the climates of the world provides a strong reason for encouraging research on these problems, and is reinforced by the recognition that human activities are developing towards a scale from which significant effects on world climate might arise within the next century. . . . No purely theoretical studies of world climate can properly achieve their objectives without an adequate base in observational data. It is important that there should be not only a record of the climates of the past on all time scales, but also a continuing record of short-term fluctuations of climate (variations over a few years) coordinated on a global scale.

More data were required but the availability of quantitative data in atmospheric, hydrological and coastal processes, for example, together with the collection of information associated with national mapping programmes such as those for soil survey, produced a situation in which the need for further process studies was highly desirable. The pages of many of the new journals initiated after 1960 (Table 1.1. p. 6) were commonly occupied by results of investigations of environmental processes and existing journals also reflected an influx of process-based papers. In addition international collaboration was encouraged by the International Geographical Union by establishing commissions that included one on present day processes, and a later one on field experiments in geomorphology (Slaymaker, Dunne and Rapp, 1980).

Process in the branches of physical geography

In attempting to summarize the developments in studies of environmental processes pertinent to, and made by, physical geographers it is evident that it is extremely artificial to distinguish the contributions made by physical geographers *per se* from those contributed by practitioners of other disciplines. The distinction has to be considered at least to some extent because only by identifying what has been done in the past can one see how deficiencies remain to be corrected in the future research agenda. Thus in a review of *Models in Geography* Slaymaker (1968, p.407) concluded that:

> With the publication of this book, British geographers can no longer be content to quote the aphorism about geography being what geographers do, and then gracefully retire from the discussion as though there were nothing to add. This volume demonstrates beyond any further doubt that the traditional classificatory geographical paradigm is inadequate and that, in the context of the 'new geography', an irreversible step has been taken to push us back into the mainstream of scientific activity by way of the uncomfortable and highly specialized process of model-building.

Also throughout the branches of physical geography it was necessary to convince members of other disciplines that physical geographers had a contribution to make and this had to be done by showing what could be done rather than by simply stating in advance that a physical geographer had much to contribute.

In Britain, with the foundation of the Natural Environment Research Council in 1966, it was notable that physical geography was not mentioned in the charter of NERC (Hare, 1966) as one of the environmental sciences that the Council would support. Hare wondered whether geographers could get away with posing as being simultaneously earth scientists and social scientists and this is what did happen in the ensuing decade. Also in 1965 the Environmental Science Services Administration (ESSA) was instituted in the United States and its concern with the interactions among air, sea and earth and between the upper and lower atmosphere embraces the scope of physical geography, although a physical geographer was not significantly involved in the early stages (Hare, 1966). Contrary to (uninformed?) views sometimes expressed it is acknowledged that physical geographers are respected for the contribution that they can make – when they do make a specific contribution. The contributions that were to be offered with the study of process may be surveyed from the viewpoint of soil science and the biogeographer, from that of the climatologist and the geomorphologist and then from the field of hydrology which to some extent provided a new focus of interest for physical geographers and one that proffered a link between at least the geomorphic and climatic aspects of physical environment. In each case it is evident that physical geographers had to become conversant with progress in related disciplines and a number of excellent papers in *Progress in Physical Geography* (1977—) provide reports reflecting progress in other disciplines.

In the study of soils it has been argued (Bridges, 1981) that soil geography aims to 'record and explain the development and distribution of soils on the surface of the earth'. As such it is located between geography and soil science, which Bridges visualizes as concerned more with the biology, chemistry and physics of soil all of which converge on the study of soil fertility and crop production. However this definition firmly identifies the contribution of physical geographers to the investigation of soils as primarily a historical or global one and that is not necessarily unanimously accepted in all the books produced by physical geographers in the 1970s (Table 5.1).

Table 5.1: Stated purposes for the physical geographer contributing to the study of soils

Date	Authority	Objective	Source
1981	E.M. BRIDGES	'The aim of soil geography is to record and explain the development and distribution of soils on the surface of the earth.'	*Progress in Physical Geography* 1981
1981	E.M. BRIDGES and D.A. DAVIDSON	'The aim of soil geography is to record and explain the development and distribution of soils on the earth's surface. It is a branch of learning which lies between soil science and geography and is of particular importance to both subjects. . . . soil studies are one means of integrating large parts of physical geography in a way which is of immediate relevance to mankind and in this way gives coherence to geography as a whole.'	*Principles and Applications of Soil Geography* London: Longman
1981	A.J. GERRARD	'Modern research is increasingly demonstrating the close dependence of soils and landforms,and a new discipline "soil geomorphology" or pedogeomorphology as proposed by Conacher and Dalrymple (1977) seems to be emerging, incorporating traditional approaches to soils as well as modern soil engineering.'	*Soils and Landforms* London: George Allen & Unwin
1972	J.G. CRUICKSHANK	'. . . soil geography is largely the study of the spatial variation of the interaction of environmental elements; and this is how soil is made.'	*Soil Geography* Newton Abbot; David & Charles

Increased attention devoted to processes was first exemplified by measurements of the variables involved in the soil system including soil moisture, soil organic matter content, soil pH, and this then necessitated knowledge of the measurement techniques themselves. Once process variables had been isolated and quantified then the interaction of dynamic process variables with spatially distributed variability of soil properties could be investigated (Trudgill, 1983). Although attention to processes at this level could be proceeding to what some geographers would clearly categorize as soil science, the movement did catalyse the development of a more three-dimensional view of the soil. Thus Runge (1973) interpreted soil properties (s) as functions of organic matter production (o), water for leaching (w) and time (t) in a form $s = f(o,w,t)$ where all parameters are fairly easily measured. To achieve a more three-dimensional model of soil it was necessary to proceed beyond the catena concept that had been introduced into soil studies based on work in East Africa, and development depended upon close relation between soils and the landsurface. One of the most striking developments was a nine-unit landsurface model proposed (Conacher and Dalrymple, 1977) as an appropriate framework for pedogeomorphic research. This model developed from one that was earlier suggested in relation to slope processes (Dalrymple, Conacher and Blong,

1969) and was important because it introduced the term *landsurface catena* to refer to a three-dimensional slope extending from interfluve to valley bottom and from the soil/air interface to the base of the soil, with arbitrary lateral dimensions. Each landsurface catena is composed of *landsurface units* which are identified and defined according to responses to single or groups of contemporary geomorphic (in 1968) and also to pedologic (in 1977) processes. The processes include interactions amongst soil materials, water and gravity or the mobilization, translocation and redeposition of materials by water flow and mass movements, whilst the responses are identifiable physical and morphological properties of the soil and also of surface morphology. Recognition of aperiodic soil/water/gravity events in a landsurface catenary concept makes this approach appropriate for studying the dynamic interaction between soil properties and landforms. Although Ruellan (1971) distinguished between those researchers who attach great importance to the geomorphological processes of erosion and deposition (allochthonists) and those who attribute the major characteristics of soils to pedological processes (autochthonists), Gerrard (1981) in his treatment of soils and geomorphology argues that soils are the results of the interaction of both sets of processes.

One subject studied relatively little by physical geographers has been the occurrence of soil erosion. Although there have been some investigations of a functional kind relating soil erosion amount to controlling variables in areas like Zimbabwe (Stocking, 1977), studies of soil loss are potentially very useful (Stocking, 1980) and may be undertaken by detailed process investigations which are usually concerned with parts of the erosion process or with laboratory measurements (e.g. De Ploey, 1983), by empirical investigations which monitor output in relation to input and use a relation similar to the Universal Soil Loss equation; and factorial survey methods which Stocking (1980) visualizes as analysis and collation of the spatial pattern of all factors which relate to soil loss erosion. Although interest in soil erosion has developed in the late 1970s (e.g. Morgan, 1979) it is a paradox that earlier work, for example developing from the Universal Soil Loss Equation (Wischmeier, 1976), did not materialize.

In biogeography it is not easy to separate natural from cultural biogeography or to disentangle the contributions which have originated in ecology from the work undertaken by physical geographers. Clues to the foci of several biogeographers may be gleaned from definitions in the number of texts on biogeography that have emerged to fill a gap that was acute until 1970 (Table 5.2). Very significant was the movement by ecologists towards trophic – dynamic ecology which was developed by Lindeman in a classic paper published in 1942 which built upon earlier conceptual frameworks and treated natural ecosystems on the basis of the capacity of their primary producers (photosynthetic plants) to capture part of the incident solar and atmospheric energy, and to incorporate it into the dry organic matter that would subsequently yield it to the grazing and decay food webs. Focus upon ecosystems in this way also required emphasis upon biogeochemical cycles which are the pathways whereby mineral nutrients are cycled through the world system and this was important as a major thrust in the systems approach (chapter 7, p. 156). In an ecosystem, identification of the trophic levels within the feeding hierarchy could be necessary in studies of specific areas which could then proceed to establish 'what eats what' and therefore how the trophic structure is built up. An extremely detailed investigation was based upon the Hubbard Brook forest of New Hampshire, USA where the pathways of chemical elements were studied under natural conditions and then after deforestation the biogeochemical cycles were investigated again and during subsequent recovery (Likens *et al.*, 1977).

Table 5.2: Attitudes of biogeographers in recent texts

Date	Authority	Objective	Source
1977	N. PEARS	'. . . the biogeographer may study the same phenomena as the ecologist, but he usually places as much emphasis on the distributional aspects as on the environmental relationships in this study. Further he will tend to stress the role of man in these patterns and processes.'	*Basic Biogeography* London: Longman
1971	B. SEDDON	'. . . forces that shape the territories of living things and the problems that are posed by their geographical distribution.'	*Introduction to Biogeography* London: Duckworth
1971	J. TIVY	'. . . the geographical focus in biogeography is two-fold: first, a study of the intimate relationships between the organic and inorganic elements of the earth's environment; and secondly the reciprocal relationship between man and the biosphere.'	*Biogeography: A Study of Plants in the Ecosphere* Edinburgh: Oliver & Boyd
1971	D.A. WATTS	'. . . to interpret the differential patterns of distribution among organisms and their changing relationships with each other and their environment both in time and space.'	*Principles of Biogeography* Maidenhead: McGraw Hill
1979	I.G. SIMMONS	'Biogeography . . . to describe our study of the biosphere and of man's effects on its plants and animals and the ecological systems of which they are a part.'	*Biogeography: Natural and Cultural* London: Arnold
1983	P. FURLEY and W. NEWEY	'Biogeography is a field of study which has come to assume a slightly different meaning for different disciplines. Literally it may be defined as "the geography of living organisms" or since geography is traditionally the study of living things and particularly man in relation to the environment, the most accurate job description of the subject may be the "geography of the biosphere".'	*Geography of the Biosphere* London: Butterworths

Studies of climatology embraced process investigations in at least three ways, each corresponding to a particular scale. At the global scale numerical modelling has developed and was applied to the planetary boundary layer, facilitated by the development of powerful computers used in short-range forecasting research, and then in the construction of general circulation models and thence models of global climate. At the meso scale models of synoptic and planetary scale motions are constructed for both research and forecasting purposes (Atkinson, 1983) and involve general models of meso-scale flows involving the planetary boundary layer (PBL) and models of particular kinds of meso scale circulation such as sea/land breeze models, urban circulation models, and slope, mountain and valley wind models. Also at this scale has been research on precipitation areas which has shown that the vertical air motion responsible and the areas themselves are organized into a hierarchy according to the horizontal size. Atkinson (1978) has reviewed the small meso-scale precipitation areas (SMPA) and large meso-scale precipitation areas (LMPA) and the way in which these areas are associated with frontal systems. Although the SMPAs last for 3–4 hours and move parallel to and ahead of the front, a theoretical explanation for the size of SMPAs is still required.

At the most detailed scale, research developments have embraced microclimatology and also applied developments such as agricultural meteorology. Microclimatology was stimulated by Geiger's *The Climate near the Ground* (Geiger, 1965).

It has been suggested that, because physical geographers have traditionally been involved in atmospheric processes, hydrology and the soil, it is logical that contributions could be made in the characterization of the plant environment both spatially and temporally (Hanna, 1983). In this field as in others physical geographers are only one group of contributors to research but recent endeavours have included field measurement of soil water and relationship between condition of the environment including soil water status and plant growth, and it is suggested that the impact of atmospheric pollution on crop growth and yields in advanced industrial countries is an area of research where physical geographers could make a significant contribution.

In geomorphology the impact of studies of process was perhaps most substantial and also the most dramatic in physical geography. At least six significant antecedents may be discussed for the study of geomorphological processes and the *first* was undoubtedly the work of Grove Karl Gilbert. In his work in the western part of the United States not only did he describe physical erosive processes but was also able to derive a system of laws governing progress from initial to adjusted forms (Chorley, Dunn and Beckinsale, 1964) and in his *Report on the Geology of the Henry Mountains* (1877) he provided the first major treatment by a geologist of the mechanics of fluvial processes. In 1914 he published a remarkable investigation into *The Transportation of Debris by Running Water* and this included the results of laboratory experiments. Gilbert is now acknowledged to have been a brilliant geomorphologist who made a contribution which anticipated many of the developments half a century later and whose deductions regarding stream and landscape mechanics 'have given new life to quantitative geomorphology in the twentieth century' (Chorley, Dunn and Beckinsale, 1964, p. 572). A *second* antecedent could be found in work by engineers, and R.A. Bagnold published his monumental *Physics of Blown Sand and Desert Dunes* in 1941, in which he stipulated the bases for the underlying processes in desert areas. Bagnold subsequently worked on processes involving fluids other than air and contributed to understanding of beach formation by waves based upon wave tank experiments (Bagnold, 1940) and also to the analysis of fluvial processes (Bagnold, 1960) and in one of his last papers (Bagnold, 1979) he reviewed fluid flow in general. Work by other engineers was obviously germane to the investigation of geomorphological processes but was not fully appreciated until the 1960s. A *third* source derived from Scandinavia where in 1935 F. Hjulstrøm published the results of field and laboratory investigations related to the River Fyris and identified relationships between stream velocity, particle size and the processes of erosion, transport and deposition that became of fundamental significance in sedimentology as well as in studies of geomorphologic processes. Later research by Sundborg (1956) on the river Klarälven, also in Sweden, led to some modification of the relationships produced by Hjulstrøm. Also in Scandinavia was an important study of the mass movement processes on the slopes of Kärkevagge (Rapp, 1960) and this was important not only because it endeavoured to quantify all of the processes that affect a slope in a subarctic environment, but also because it established the relative significance of the different processes and concluded that the most effective agent of removal was running water removing material in solution.

The remaining antecedents all emerged in North America but for slightly different reasons. *Fourthly* there was research directed by A.N. Strahler at Columbia University, already referred to in relation to quantitative advances (p. 56). In the Columbia school of geomorphology measurements were made of processes operating on stream channels and slopes of a number of areas (e.g. Schumm, 1956), later the focus turned to coastal processes as exemplified by changes in an equilibrium beach related to the tidal cycle (Strahler, 1966) and perhaps most significant was Strahler's advocacy of the need for a dynamic basis for geomorphology (Strahler, 1952). This significant paper endeavoured to extend geomorphology from what is now appreciated to be a functional viewpoint towards a more realist view, as indicated by the aim (Strahler, 1952, p. 923):

> . . . to outline a system of geomorphology grounded in basic principles of mechanics and fluid dynamics, that will enable geomorphic processes to be treated as manifestations of various types of shear stresses, both gravitational and molecular, acting upon any type of earth material to produce the varieties of

> strain, or failure, which we recognize as the manifold processes of weathering, erosion, transportation and deposition.

This fundamental approach was later to be followed by other realist approaches to geomorphology including Carson (1971).

A *fifth* contribution came from the ideas of dynamic equilibrium advanced by J.T. Hack (1960). Influenced by the work of G.K. Gilbert, Hack argued that the concept of dynamic equilibrium provides a more reasonable basis for the interpretation of topographic forms in an erosionally graded landscape, that every slope and stream channel in an erosional system is adjusted to every other, and when the topography is in equilibrium and erosional energy remains the same, all elements of the topography are downwasting at the same rate. In this view the accordant summits in areas like the ridge and valley province of the USA were interpreted as the inevitable result of dynamic equilibrium rather than as remnants of earlier erosion cycles, and this followed from the assumption (Hack, 1960, p. 81) that:

> It is assumed that within a single erosional system all elements of the topography are mutually adjusted so that they are downwasting at the same rate. The forms and processes are in a steady state of balance and may be considered as time-independent.

To some extent the attitudes of dynamic equilibrium and of longer-term perspectives were reconciled by the extremely significant paper (p. 161) on Time Space and Causality in Geomorphology (Schumm and Lichty, 1965).

A final and *sixth* antecedent may be found in the investigation of magnitude and frequency of geomorphic processes (Wolman and Miller, 1960) and subsequently of fluvial processes in general (Leopold, Wolman and Miller, 1964). This influential paper proceeded from an explanation that for many processes above the level of competence, the rate of movement of material can be expressed as a power function of some stress, to demonstrate that the largest portion of sediment transported by rivers is carried by flows which occur on average once or twice each year, and that transport of sand and dust by wind follows the same laws. The analogy was drawn between the efficacy of geomorphic processes and that of a giant, a man and a dwarf attempting to cut down a forest. The dwarf maintains an assault on the trees for long periods and achieves little (the equivalent of frequent, low-magnitude events); the giant sleeps most of the time but occasionally wakes and causes great destruction (a catastrophic event) whereas the man works regular hours and systematically achieves the greatest effects (events that occur once or twice each year). It was concluded that (Wolman and Miller, 1960, p. 54):

> Closer observation of many geomorphic processes is required before the relative importance of different processes and of events of differing magnitude and

> frequency in the formation of given features of the landscape can be adequately evaluated.

In 1964 publication of *Fluvial Processes in Geomorphology* (Leopold, Wolman and Miller, 1964) ushered in a new era of process investigations. This book was in effect the first which emphasized contemporary processes and underlying physical principles and it focused upon river channels, drainage systems, slopes and had some reference to climatic-inspired systems. In their introduction the authors explained that (Leopold, Wolman and Miller, p. 7):

> At the present deductions are subject to considerable doubt, for the detailed properties of landform have not been studied carefully enough and the fundamental aspects of most geomorphic processes are still poorly understood. So long as this is true, the interpretation of geomorphic history rests on an exceedingly unstable base.
>
> Accordingly, we plan to concentrate on geomorphic processes. The emphasis is primarily upon river and slope processes. . . .
>
> Process implies mechanics, that is, the explanation of the inner workings of a process through the application of physical and chemical principles;

and the import of this publication was reflected in the citations that have subsequently been made to it and in the reviews which appeared including: Glenn W. Frank writing in *Professional Geographer* (1965, 17, p. 46):

> 'This book is the most penetrating and comprehensive book dealing with geomorphological processes available today.'

B.W. Sparks in *Geographical Journal* (1965, 131, p. 133):

> 'If students are to be well grounded in a healthy and promising attitude to fluvial geomorphology, they will have to grasp the ideas contained in this book.'

E.H. Brown in *Geography* (1965, 50 p. 193):

> 'Geomorphology is changing rapidly; this book signposts one of the ways in which it will surely develop.'

In the 1960s and subsequently the discovery of the need to study processes led to appreciation of the need for familiarity with subject matter from interdisciplinary origins and to acquire and develop techniques that could be employed for process investigations and particularly for empirical measurements. Thus research papers were sometimes apparently preoccupied with technique simply because techniques had to be defined and refined for the purpose of the physical geographer. Just as the scope of physical geography had become unbalanced (Brown, 1975), so the content of geomorphology now began to show signs of imbalance as the emphasis moved towards fluvial geomorphology and hydrology.

One area where the importance of process measurements had always been appreciated was in limestone areas and these were the focus of many of the earliest, and simplest process techniques developments. Initial progress was made in tracing water discharged at resurgences but this was succeeded by deductions of limestone erosion rates, by determination of drainage areas, by distinction of the significance of percolation water and of conduit flow and thence of application of the results of process investigations to the derivation of modified models of landscape development as proposed by Smith and Newson (1974) for the Mendips. In coastal geomorphology the process foundation was also well established by the achievements of engineers and of research institutes and incorporated theoretical and laboratory investigations as well as empirical ones. The coastal geomorphologist therefore focused upon the processes that control shoreline equilibrium and the significance of long-shore currents and sediment transport, of wave activity in relation to the swash, nearshore and offshore zones, and to the less-studied influence of tides and of impulsive events such as tsunamis. Research investigations were exemplified by work by C.A.M. King and other staff from the University of Nottingham who investigated changes in the coast of south Lincolnshire at Gibraltar Point from 1951 to 1979. The impact of processes in the study of coastal geomorphology was reflected in the content of a major text (King, 1972) and a further text (Davies, 1973) included a map of wave environments of the world, which is one way in which geographers endeavoured to establish spatial variations.

It seems as if each major branch of geomorphology had at least one new textbook which served to disseminate techniques and ideas and to serve as a baseline from which new research and teaching could develop. In relation to deserts Cooke and Warren (1973) provided a standard *Geomorphology in Deserts* in which they were primarily concerned with desert landforms, the materials that compose them, the processes of debris preparation, erosion and transport that modify them, and the environmental factors influencing all these phenomena. In the fields of glacial and periglacial geomorphology the first modern textbook (Embleton and King, 1968) was soon followed by others (Price, 1973); and later work utilized a systems framework (Andrews, 1975; Sugden and John, 1976) as indicated later (p. 159). In their introduction Sugden and John indicated in 1976 (p. 1) that the study of glacial processes and forms had been left out in the cold and poorly understood because a gulf had arisen between those who study glaciology and those who study glacial landscape and deposits, and further that:

> Perhaps there is a need for a more glaciological type of geomorphology and a more geomorphological type of glaciology. There is now a strong case for a realistic dialogue between those studying glacier dynamics and those studying forms. Until this occurs, there can be few spectacular advances such as those achieved recently in fluvial and slope geomorphology.

The way in which glacial processes were studied was by investigations in areas including Iceland, Baffin Island, Antarctica, Alaska, Scandinavia, and the European Alps and it was from such research that the understanding and interpretation of past glacial systems was enhanced, as in Scotland by Sissons (1967). Thus glacial geomorphology was defined (Gjessing, 1978) as concerned with bedrock forms and superficial deposits produced by glacial and fluvioglacial processes in areas of present glaciers as well as in areas covered by glaciers during the Quaternary. In periglacial geomorphology, knowledge of contemporary processes was achieved by investigations of permafrost in Siberia and in Alaska and Canada and of cryonival processes in areas such as Spitsbergen. Thus work, such as that on thermokarst in Siberia in relation to the development of lowland relief (Czudek and Demek, 1970), had a significant influence on research in areas of Quaternary periglacial morphogenesis. The influence of theoretical research in glaciology such as that by the physicist J.F. Nye (1952) was substantial. Nye derived equations for glacier flow assuming that ice is a perfectly plastic substance, that it flows down a valley of constant slope, and that the conditions of temperature, accumulation and ablation are simple and uniform, and his model could be compared with field observations. Textbooks in periglacial (e.g. Davies, 1969; Pewé, 1969; Washburn, 1973), as well as textbooks in glacial geomorphology acknowledged the contributions made from other disciplines, from research institutes and from international conferences. Thus the Institute of Arctic and Alpine Research (INSTAR) at Colorado founded in 1951 in the United States, the Scott Polar Research Institute in Cambridge, UK (1920—) and the V.A. Obruchev Institute for Permafrost Science in the USSR (1930—) are examples of research institutes that contributed significantly to developments in the science of cold regions.

Slopes had become the focus for revived interest since 1950 (Strahler, 1950b) and had been the subject for quantitative description and analysis (Bakker and Le Heux, 1952) and then for measurements which could lead to measurements in specific areas using for example the Young Pit (Young, 1960) and providing many indications of rates of erosion (A. Young, 1974). Slope process investigations were influenced by empirical measurements first using single, and later continuous, recording; by theoretical approaches which were admirably exemplified in the book *Hillslope Form and Process* (Carson and Kirkby, 1972); and by stability analysis which utilized the factor of safety approach and related approaches used by the civil engineer. These three strands tended to complement the well established qualitative models of Davis and Penck and the blend of the four approaches adopted in textbooks tended to vary according to author and date of writing, with an emphasis on factors and measurement in the earlier books (e.g. Young, 1972) and a more theoretical foundation stressed in others (Carson and Kirkby, 1972).

As noted above, Sugden and John (1976) implied that spectacular advances had occurred in slope and in fluvial geomorphology. In the latter some 27.7 per cent of British research could be categorized as fluvial including both processes and landform development in 1975 (Gregory,

1978d) but the pattern was complicated because of interaction with hydrology. From a geomorphological viewpoint the great increase in research investigations particularly since Horton's 1945 paper and the research of the Columbia school has been reviewed and ascribed (Gregory, 1976) to research associated with seven contributing themes, namely network morphometry, drainage basin characteristics, hydraulic geometry, river channel patterns, theoretical approaches, dynamic contributing areas, and the palaeohydrology – river metamorphosis approach. The latter owed much to research by Schumm and his students and is reviewed further in chapter 8 (p. 172) whereas the other themes owed much to increasing knowledge of, and dependence upon, hydrology. In addition to many research contributions in the field of fluvial morphology which are reflected in the books that succeeded Leopold, Wolman and Miller (1964) and were devoted to streams (Morisawa, 1968), to drainage basin form and process (Gregory and Walling, 1973), to water and environmental planning (Dunne and Leopold, 1978), and to alluvial river channels (Richards, 1982), there were also books produced by geographers dealing with hydrology (Ward, 1967) and with aspects of hydrology such as floods (Ward, 1978). Indeed, although some physical geographers perceive hydrology to be a separate field of scientific enquiry and, for example, *Progress in Physical Geography* maintains progress reports on fluvial geomorphology as well as on hydrology, other physical geographers have advocated the notion of geographical hydrology. Thus Ward (1979) uses the term geographical hydrology to connote the hydrology worked on by geographers and noted that:

> the engineer resorts to empiricism and coefficients, to simplification and generalization of systems and processes. The geographer on the other hand, is primarily interested in how the landscape works, and in man's interactions with it, and thus recognizes that water is but one of the terrestrial phenomena in the total complex interacting ecosystem in which he is really interested. This implies that much of the geographer's hydrological endeavour is directed not towards the solution of a specific hydrological problem but towards a more complete understanding of the landscape. . . . Recent trends in both disciplines have begun to blur such distinction.

Ward (1979) noted with alarm the signs of what he called 'Frontier man's fright' whereby after working in an area for some time without apparent response from professional colleagues in related disciplines there is a tendency to desert it just as its true worth is to be recognized by others:

> . . . . within hydrology at the moment we can see a growing interest among civil engineering hydrologists in 'process' studies at a time when some geographers are beginning to show signs of impatience with these (Ward, 1979, p. 393).

It is debatable whether the idea of prefixing other disciplines with the word geographical is indeed a sound one because it may tend to suggest a greater separation of physical geography endeavour from that of other researchers when in fact there is no clear distinction in methods or objectives. In addition to the seven types of investigations specified above as referring to fluvial geomorphology, the main focus of research by physical geographers was on small instrumented areas as indicated below (p. 107) and this developed towards the assessment of sediment and solute yields, as exemplified by Walling (1983) and to the assessment of human impact including that on urban areas (e.g. Hollis, 1979). The contribution now made by physical geographers to international cooperation and research is exemplified in the work of D.E. Walling, first as Secretary and then as President, of the Commission on Continental Erosion of the International Association of Hydrological Sciences. Physical geographers have made substantial contributions to hydrology and this has arisen not only from contributions concerned with the drainage basin and with run off generation but also from hydrometeorology.

The above review of the branches of geomorphology would perhaps indicate that the increasing emphasis upon process was accompanied by increasing fragmentation of geomorphology as the branches each became more closely associated with other disciplines. In a review of physical geography Marcus (1979) has noted that in addition to maintaining some contact with human geography, 'most physical geographers today keep one foot in the AAG and the other in a cognate scientific society. It is a reality of our professional lives.' However although fragmentation was a feature evident in the 1960s, in the 1970s there were at least two trends that tended to foster integration: namely the use of common techniques and a more realist approach. Techniques were initially to some extent specific to particular branches of geomorphology but increasingly were found to have much in common and this trend should be further intensified as the potential of environmental monitoring by remote sensing becomes even more widely available and fully utilizes the opportunities offered by developments in microelectronics. An early book on techniques in geomorphology (King, 1966) covered the range then available, specific manuals have been produced such as the US Geological Survey *Techniques of Water Resources Investigations* and the *Technical Bulletins* of the British Geomorphological Research Group, and an edited manual of geomorphological techniques (Goudie, 1981a) embraces the earlier developments. Although techniques assumed a more significant role, as indicated by these publications and by major sections in textbooks (e.g. Gregory and Walling, 1973), specific laboratory and field techniques courses were not quickly adopted in teaching physical geography in all departments although a self-paced course was one that had been developed (Clark and Gregory, 1982).

A more realist approach had been advocated by Strahler (1952) and had the implication that not only would there be an approach to processes in terms of physical principles, but that this should also involve a similar approach to materials. After Strahler one of the earliest movements in this direction was by E. Yatsu in *Rock Control in Geomorphology*, which proposed a quantitative approach to the underlying mechanics of processes, and as Yatsu (1966) remarked:

> Geomorphologists have been trying to answer the *what*, *where* and *when* of things, but they have seldom tried to ask *how*. And they have never asked *why*. It is a great mystery why they have never asked *why*.

Subsequently, in the 1970s this approach was also addressed by M.A. Carson in *The Mechanics of Erosion* (Carson, 1971) in a book which provided a unified introduction to the mechanics of erosional processes and which was aimed at undergraduates in earth science. In this important book Carson considered the concept of stress, the mechanics of fluid erosion, stress–strain–strength interrelationships, mass movements in rock and soil masses, and mechanics of glacial erosion and concluded that (Carson, 1971, p. 166):

> The neglect of the study of geomorphic processes by workers in geomorphology must rank as one of the most puzzling features in the development of this discipline. There are those who put the blame for this on the writings of W.M. Davis, arguing that his all-embracing model of landscape evolution left little for geomorphologists to do besides applying the model successively to new areas. And yet this is hardly fair. Two decades ago, Strahler tried valiantly to infuse a genetic approach into the subject but was greeted with meagre response. . . . There is something strange about a subject in which its research workers are willing to dabble at the application of the jargon of thermodynamics but unwilling to apply even the most rudimentary aspects of mechanics to these problems. One suspects that geomorphology will emerge as a reputable discipline only when its students have become well versed in the established principles of natural science.

Carson (1971) identified the achievement that had already been made by the mechanics approach and predicted that more would follow, but that because most erosional processes are weathering-limited to varying degrees, therefore on a geological time-scale erosion mechanics must be closely linked with the mechanics and chemistry of rock breakdown. However, progress in a similar way towards an understanding of the mechanics of weathering processes (e.g. Curtis, 1976) requires not only an adequate knowledge of chemistry and of exchange reactions, but also of processes at a different scale. Geomorphology has been categorized (Chorley, 1978) as a meso-scale science and as in the case of the other earth sciences the extent to which it is realistic to extend geomorphological processes to the micro-scale is arguable, especially as the positivist view is no longer sustained in physics and other natural sciences. There can be little doubt, however, that geomorphology within physical geography needs to overcome its unnatural reluctance to become familiar with the principles underlying landscape processes because this will not necessarily prejudice the retention of a meso-scale approach and it has been achieved by sedimentology (e.g. Allen, 1970) and by soil science. For example,

Carson (1971, p. 167) argued that an approach more firmly based in mechanics would lead to a more integrated understanding of geomorphological processes so that the student would be more impressed by unity rather than alarmed by superficial diversity. We have yet to reach the position where sufficient students appreciate the import of underlying theory and this is why the geomorphologist reaches for his soil augur once theory is mentioned (Chorley, 1978) and perhaps the situation will be resolved only when training in physical geography demands greater familiarity with scientific and mathematical methods and notation. S. Gregory (1978) underlined the need for a more secure mathematical and scientific foundation for physical geographers.

One way in which this has been achieved is by devising larger units for education in which physical geography is more closely allied with other earth and environmental sciences as noted by Clayton (1980a). At pre-university level the originally dominant historically-based approach has been replaced by teaching which often focuses upon the impact of human activity and upon management considerations but gives much less emphasis to the mechanics and principles of landscape function. This is rather like putting the cart before the horse and it is very difficult later to take up the study of the horse when all previous emphasis has been placed upon the cart! Physical geography in general, and geomorphology in particular, could be in danger of developing into a science which does not even attempt understanding of such basic mechanics and principles, partly by analogy with the decline of positivism in human geography. This has continued to be the fault of the superficial application of the effects of human activity and of the systems approach (see p. 116, and 140). Despite the appearance of other books (e.g. Statham, 1977) that follow the approach used by Carson and also of excellent illustrations of research benefits to be realized by this method (e.g. Prior, 1977) there is still considerable potential for development in this direction. Although some physical geographers such as Thomas (1980 p. v) have argued that study of the energetics of the land surface 'has perhaps robbed the subject of some of its scope and depth' it can be argued that investigations of process energetics have provided some depth of understanding and have extended the scope towards recent temporal change (p. 153).

Methods of process investigations

When process investigations began to increase in the 1960s it was not realized exactly how difficult they would prove to be. The primary objective is to measure energy transfer in physical geography and the closer one gets to the transfer the more realist is the approach, in contrast to the functional which tends to employ surrogates for process work functions. The fundamental problem confronting process measurement is that there is energy transfer taking place at what is effectively an infinite number of points in space and in time. To measure or assess processes there has to be a strategy adopted for sampling from what is an infinite population. In geomorphological terms Church (1980) has expressed this situation in terms of mass or energy fluxes or changes in content of a storage or control volume. Thus a sequence of observations will specify magnitudes of fluxes (q_t) or successive states of a system parameter (x_t). The net amount of work which occurs is often estimated by observing the frequency of an externally applied effective stress (such as a weather- or hydrologically-related stress) which is more easily measured or is accessible from records. Such a stress is converted into the sequence of interest via a rating calibration or transfer function which links the applied stress and material flux or strain. Change in landscape over a period $T = t_2 - t_1$ can be measured either by integration of fluxes

$$\int_{t_1}^{t_2} q_t \, dt$$

or by net displacement of a system parameter ($x_{t_2} - x_{t_1}$). Church (1980) clearly identifies the fact that the character of geophysical event sequences complicates simple analysis. Three features which are significant are trend, which includes well defined cyclic behaviour; persistence, when a particular value in a sequence is constrained by adjacent values; and intermittency, whereby there is a tendency for non-periodic grouping of like values over long periods of time, which effect is called the 'Hurst phenomenon'. These complications are illustrated in Figure 5.1 and an example of variance spectra also conjectured by Church (1980) is also illustrated in Figure 5.1 and introduces a theme which will be resumed in chapter 8 (p. 161).

Four main strategies have been utilized to provide the data for analysis via a sequence such as that indicated in Figure 3.1. First are theoretical approaches which depend upon some empirical knowledge to apply theoretical concepts such as the continuity equation. Secondly are direct empirical measurements, thirdly are experimental investigations; and fourthly there are historical techniques.

Direct empirical measurements are provided by national monitoring agencies including measurements of meteorological elements and of river discharge. In many cases such measurements may not be available with spatial and temporal sampling frequencies that meet the demands of research programmes. Hence additional measurements have often had to be made and not only should these be related to an *a priori* hypothesis but also they may be derived from small experimental areas. Although measurements in small watersheds were undertaken in the late nineteenth century, it was after the mid twentieth century that the movement developed very rapidly and absorbed the attention of some physical geographers. The Vigil Network scheme was initiated in the mid 1960s and involved the careful measurement of processes at individual sites and in selected study

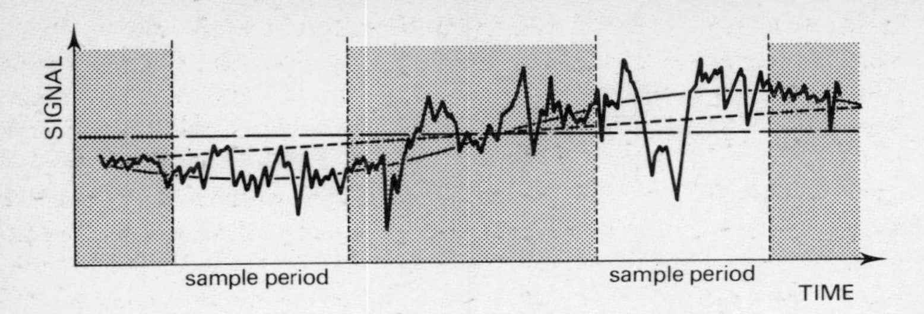

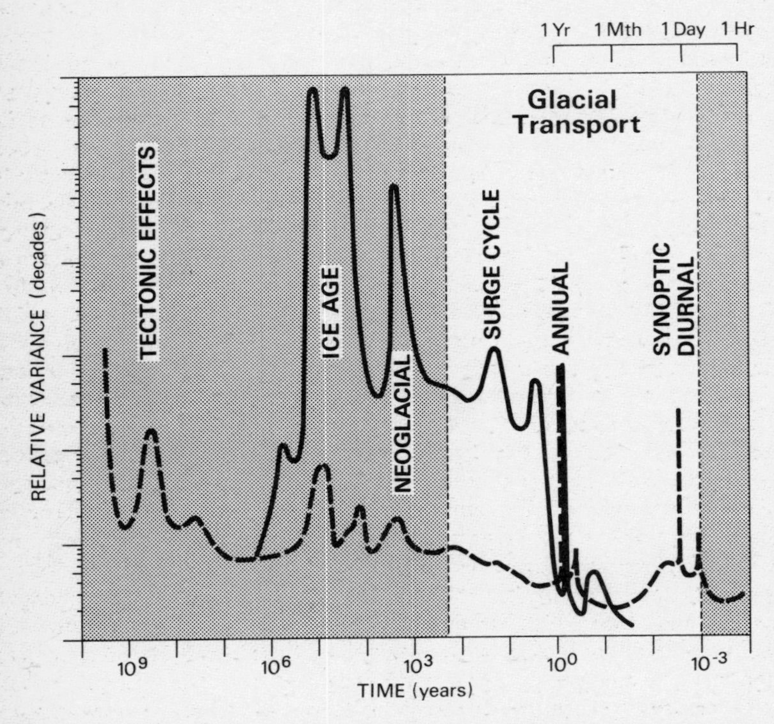

Figure 5.1: Event sequence (above) and conjectural variance spectrum (below) after Church (1980). In the event sequence is shown the long-term mean, the linear trend, the well-defined cycle, and the intermittent signal so that a typical sample period as shown will produce statistics which are biased with reapect to the whole sequence. The conjectural variance spectra for glacial transport is standardized on the annual cycle.

basins. Observations were intended to be made of stream channel changes, mass movement, vegetation changes, precipitation, reservoir sedimentation, and dendrochronology, and the intention to maintain measurements over a decade or more has been realized at some sites (e.g. Leopold and Emmett, 1965). A range of objectives for watershed experiments was identified by Ward (1971) and these extend from specific black box approaches to comprehensive studies in which there is an attempt to monitor many of the processes operating within a small area. The Hubbard Brook experimental basin (15.6 ha) in New Hampshire was monitored in great detail including the effects after complete clearing of the woody vegetation and treatment with herbicides for three years. Some small instrumented areas have been selected as representative of particular conditions whereas others were experimental, in which change due to logging or other forms of land use change could be monitored either by change in one basin or by comparison of several areas in which changes were taking place to differing degrees. A substantial number of small instrumented areas have been instituted in different parts of the world and physical geographers contributed to the initiation and running of a considerable number of these. The aims and objectives of all experiments were not always formulated as clearly as possible and there was sometimes a tendency for instrumentation to be set up in the hope that a specific problem would arise once measurements were obtained. In this and other cases a number of problems often arose reflecting lack of control and replicability of measurements, insufficient representativeness, unreliable accuracy of data, and problems of finding suitable methods of analysis of the data collected. Thus Ackermann (1966) writing about the United States suggested that 'small experimental watersheds have cost this country uncounted millions of dollars and unfortunately have yielded a small return on the investment of time and money.' However a considerable number of advances were made in the framework of small instrumented areas and there is no immediately obvious alternative to the approach when large numbers of empirical values are required. Newson (1979) has argued that process studies in geomorphology seek to quantify three systems; the mechanical sediment system, the chemical system, and the hydrological system, and it is notable that progress towards the quantification of these systems has been achieved by greater cooperation of physical geographers with scientists in other disciplines and in response to requirements of engineers and planners for resource information.

Experimental investigations embrace a range of approaches which include field plot experiments, through laboratory hardware models which attempt to use scaled-down versions of the real world, to analogue models which employ a different medium for investigation. A Kaolin glacier was an example of the last-named (Lewis and Miller, 1955) but much more extensively used have been measurements using rainfall simulators especially in relation to erosion experiments, flumes, wave tanks and wind tunnels. In all cases the potential available has probably not been fully explored in geomorphology (Mosley and Zimpfer, 1978) but of all the limitations the difficulty of overcoming the scale problem and of relating the observations to geophysical event sequences (e.g. Fig. 5.1) have been most evident.

Historical techniques in physical geography offer some of the most imaginative available and have been reviewed by the cooperation of a geomorphologist and a historical geographer (Hooke and Kain, 1982). They contend that:

> . . . evidence from the past has considerable potential for providing longer term perspectives for studies of current physical processes, for understanding the nature and causes of change and, above all, for understanding the magnitude of the impact of human activities on the physical environment.

They review graphical, written and oral sources, statistical sources and series, and non-documentary sources and clearly demonstrate the amount of scope which remains for the use of such sources. The methods available are constantly increasing in number and their utility is greater as the complexity of contemporary processes is revealed. One somewhat unconventional data source is the newspaper for reconstructing past event-frequency and character and a series of studies by M.G. Pearson (e.g. Pearson, 1978) analysed the incidence of eighteenth and nineteenth-century weather conditions in Scotland based upon reports in newspapers. This source has been reviewed by Gregory and Williams (1981), who produced maps of hazard distribution in Derbyshire 1879–1978 based upon reports in the *Derby Evening Telegraph*.

Patterns of processes

A recurrent problem encountered in investigations of processes is that measurements and analysis are conducted for a detailed level near the climatological station, in the soil pit or on an erosion plot, whereas many potential applications of process studies depend upon knowledge at a much greater spatial scale and possibly that of the world level. Study at such world levels can now be achieved first because enough process data have been collected to allow some degree of national and world pattern to be discerned; secondly because international cooperation has been initiated specifically to remedy the past deficiencies of data coverage; and thirdly because the opportunities of remote sensing are now beginning to provide further clues to the world picture.

World spatial systems afford an enormous scope for review but perhaps the dominant trend has been towards the differentiation of the earth's surface on a more realist dynamic basis to replace a more static arid functional treatment that had been current previously. This trend is exemplified by all branches of physical geography. Climatology is probably the most data-rich branch, in which maps showing mean values of climatic elements such as precipitation or temperature had been long-established and numerous attempts had been made to classify climates upon a world basis. Whereas many such attempts embodied a somewhat static view of the pressure distribution, this was succeeded by what Barry and Perry (1973) termed a kinematic view of the weather map in which the synoptic weather map is viewed in terms of airflow and the movement of pressure systems. This included what Court (1957) described as pressure field climatology and led to a range of approaches which included weather patterns and air mass climatology, all of which provided a more effective reflection of processes. A more dynamic approach of this kind was possible with the advent of computers which could analyse the necessary amounts of synoptic data. Then with the utilization of satellites when satellite climatology ushered in a new era which Barrett (1974) has characterized as providing observing systems of the earth and atmosphere, as highly convenient data collection platforms, and as connection links between widely spaced ground stations between which large daily exchanges of weather data must take place. Although much had been achieved by 1974 Barrett argued that 'the best is yet to be' (see also chapter 10, p. 211).

Also in the field of biogeography world maps of plant formations had been utilized for many years and had indeed been adopted as the basis for some climatic classification by attempting to fit the classification to the plant distribution. However, the static quality of such maps could be replaced when research on ecological energetics, or nutrient cycling and on population dynamics led towards greater use of net primary productivity (NPP), which is the material actually available for harvest by animals and for decomposition by the soil fauna and flora or their aquatic equivalents. The rate of accumulation of biomass or NPP is expressed as weight of living matter/unit area/ unit time and recent work by the International Biological Programme (IBP) has provided more accurate estimates of NPP on a world scale for both continents and oceans, and this makes it possible to rank biome types according to NPP or present processes. This offers an additional dimension to biogeography (Simmons, 1979a) and because it is infinitely renewable but subject to substantial modification by man, the NPP of an area can be viewed as what Eyre (1978) characterized as 'the real wealth of nations'.

Soil geography has also had a well established legacy of soil maps which have attempted to relate the one extreme of detailed soil survey to the other of a world distribution. Emphasis has been more towards the local scale because of the demand for soil maps in relation to agricultural and other land use purposes. Perhaps emphasis at this level has been upon soil evolution rather than upon soil dynamics, which has been treated in relation to land capability. However the increasingly similar basis underlying national soil maps allows correlation to take place more easily and a soil map of the world which is in an advanced state of preparation is envisaged (Bridges,

1978a) as the basis of soil geography in future years. In geomorphology, as in soil geography, it has not been easy to achieve representations of global spatial patterns based upon processes to succeed the long-established static morphological or structured landform regions. However, coastal environments have been classified according to wave energy by Davies (1973) and this can assist a more meaningful correlation between wave type and coastal morphology. It is in regard to erosion rates that perhaps the most dramatic progress has been made and in addition to world patterns of discharge, Walling and Webb (1983) have now reviewed earlier attempts to portray world patterns of sediment yield, some of which conflicted quite significantly, and have provided a revised map of global sediment yields based upon data from nearly 1500 measuring stations and pertinent to sediment yields from basins of 1000 to 10,000 km^2 in area. Such progress exemplifies the way in which a physical geographer can contribute by deriving a world pattern that depends upon the most recently reviewed methods for assessing sediment yield and presenting it in a form that is pertinent to applications in a wide range of disciplines.

Walling and Webb (1983, p. 95) point to the need for closer cooperation between limnologists and fluvial geomorphologists which could lead to greater understanding of sediment yield. Whereas examples from each branch of physical geography indicate the way in which spatial patterns reflecting process and dynamics can be derived, there is also the possibility of a more integrated approach. Climatic geomorphology was introduced earlier (p. 70) in relation to landscape chronology and one of the earliest approaches by Peltier (1950) endeavoured to relate climate to geomorphological process employing a semi-quantitative basis involving mean annual temperature and mean annual precipitation. In a later paper Peltier (1975, p. 129) contended that

> Climatic geomorphology is one part of a broader structure of regional geomorphology which includes both tectonic geomorphology and the stratigraphy of continental deposits.

He produced a general landform equation in which landform (LF) could be viewed as a function of geologic material (m); rate of change of geological material, structural factor (dm/at); rate of erosion (de/dt); the rate of uplift (du/dt); and the total time of duration of the process (t) in the form LF = f (m, dm/dt, de/dt, du/dt, t), and he then proposed expressions for an erosion factor (de/dt). It is certainly desirable that relationships between geomorphological processes and climatic parameters are established using the results of process studies and then extended towards the global scale, and such relationships could enhance the understanding involved in climatic geomorphology.

However it should also be possible to coordinate progress on soil and ecosystem patterns as well and a leading paper on this subject (Hare, 1973) begins with a quotation from Budyko and Gerasimov (1961):

> . . . the heat and water balance of the earth's surface is, as a rule, the main mechanism that determines the intensity and character of all the other forms of exchange of energy and matter between . . . the climatic, hydrologic, self-forming, biologic and other phenomena occurring on the earth's surface. . . .

Hare (1973, p. 171) wisely avows that 'synthesis is easy to announce, but hard to pull off' but in leading towards a focus upon energy (see p. 155) investigates the links between climate and soil, plant and animal life and concluded in 1973 that this was a more promising avenue of exploration than links with geomorphology. A considerable amount of research on energy-based and energy-budget climatology had been undertaken since 1956 when Budyko (translated 1958) attempted a heat balance of the earth's surface and proceeded from equations of

$R = H + LE$ and $P = N + E$

where

H = convective heat (enthalpy) flux
E = evapotranspiration
N = runoff
P = precipitation
R = net radiation or radiation balance defined by
$R = I(1 - a) + R\downarrow - R\uparrow$
where I = incident global solar radiation on a horizontal surface
a = albedo

$R\downarrow$ and $R\uparrow$ are the downward and upward long-wave radiative fluxes at the surface.

Budyko then proceeded to attempt a physio-geographical zonation of the earth and his approach has subsequently been developed by systems in which energy and moisture regimes are related to vegetation types (e.g. Grigor'yev, 1961), to genetic soil types (e.g. Gerasimov, 1961) and later to geomorphic zonality (Ye Grishankov, 1973). It is curious that geomorphologists have not generally explored the potentially fruitful links with energy-based climatology that are identified by Hare, who comments (1973, p. 188):

> In spite of the spread of quantitative, physically inspired methods, and a growing preoccupation with process, geomorphologists have not put their discipline on an energy balance basis to nearly the same extent, and probably with good reason.

The reasons adduced by Hare include the facts that fluvial processes tend to be dominated by extreme events rather than balance relationships, so that stochastic methods and extreme-value theory are closer to the reality of geomorphic processes than is energy-balance climatology; and that the geomorphic time scale is longer than that utilized by the climatologist. However, the climatic influence has been the subject of review on aspects of geomorphology (Derbyshire, 1976) and the map produced by Walling and

Webb (1983) is included in a volume on palaeohydrology which is one of the more recently developing temporal approaches (p. 172). Hence the prospect of an energy related and integrated physical geography including geomorphological processes may not be too far beyond the horizon. The direction

> . . . of a common, quantitative, theoretically-based language . . . in my judgement, is what physical geography needs. They lead us, also, in the direction that all science aims at, towards prediction. (Hare, 1973, p. 189).

Time and catastrophic punctuations

It is impossible to study processes in physical environment independently of time although timeless and timebound approaches have been distinguished (Chorley, 1966). For purposes of description time can be considered under four headings (Thornes and Brunsden, 1977). *Continuous* time means that observation is unceasing, such as a continuous record of river discharge; *quantized* time is when imaginary sections are used to subdivide time, as when precipitation amounts are measured daily or weekly; *discrete* time is when interest focuses upon time duration and frequency of events per unit of time; and *sampled* time is when observations can only be made at particular periods, such as weekly measurements of plant growth. Whichever heading is appropriate it is necessary to appreciate the significance of events which may occur beyond the period of normal measurement during the duration of a research programme. The significance of such events was highlighted by Wolman and Miller's (1960) consideration of the magnitude and frequency of geomorphic processes and subsequently the significance of rare events has been scrutinized. On coasts of coral atolls the effects of hurricanes are shown to be significant not only to the reef morphology but also to the organisms and the total environment (Stoddart, 1962) and in drainage basins in general and along river channels in particular the impact of rare floods and the time necessary for recovery has been investigated. In a review of previous studies it has been indicated (Gupta, 1983) that stream channels are affected by low-frequency, high-magnitude events and the persistence of such effects is greater in arid and semi-arid than in temperate areas (Wolman and Gerson, 1978). Recent research has shown how glaciers may be affected by periodic surges in which ice may be transmitted down-glacier at speeds 10–100 times faster than normal and this has been identified from small glaciers and also from some ice sheets on a continental or subcontinental scale (Sugden and John, 1976).

The incidence of such infrequent but significant events has promoted two consequences. First, there have been studies of earth hazards and a range of single and composite hazards have been identified, as indicated in Figure 5.2. However because (White, 1974, p. 3):

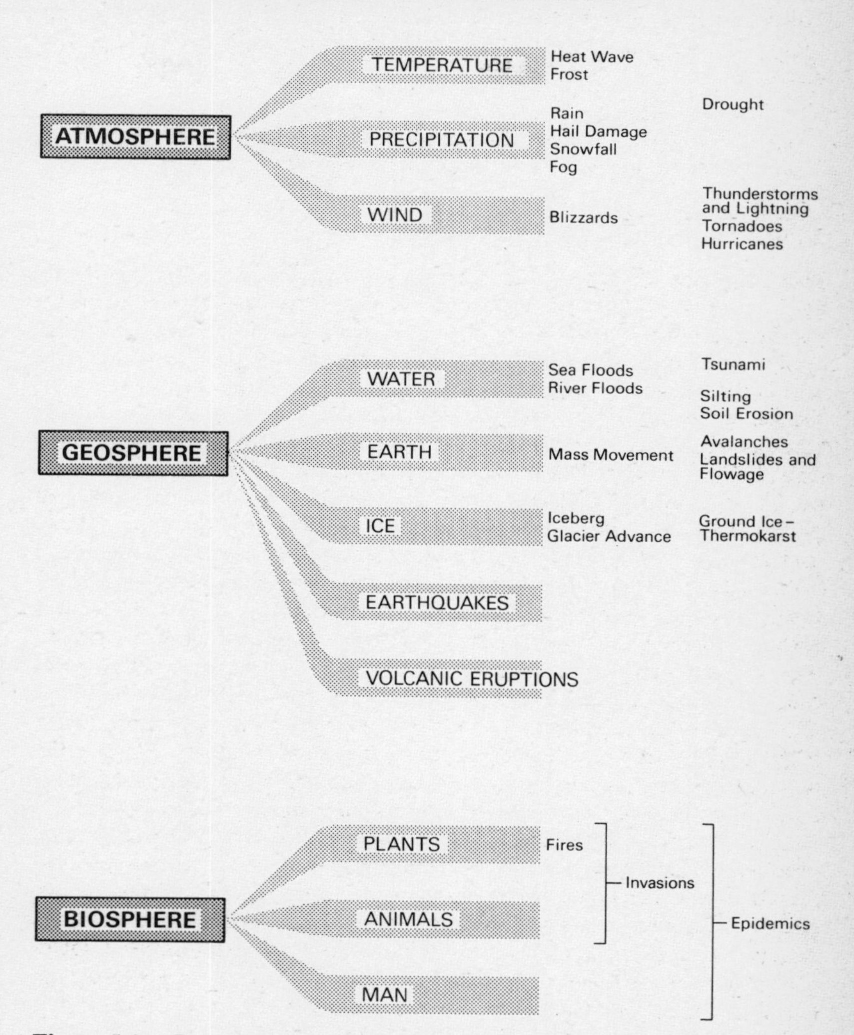

Figure 5.2: Potential earth system hazards (after Gregory, 1978d) Compound hazards are shown on the right.

> Extreme natural events illuminate one aspect of the complex process by which people interact with biological and physical systems. . . . By definition no natural hazard exists apart from human adjustment to it,

the progress in the study of hazards is reviewed later (p. 130) in relation to human activity and to applied physical geography, because as White (1973, p. 193) has also noted:

> To a remarkable degree during the 1960s, geographers turned away from certain environmental problems at the same time that colleagues in neighbouring fields discovered those issues. . . . By neglecting the theory of man–environment relationships and its applications to public policy

Second has been the revival of interest in neocatastrophism which has arisen as a concept which acknowledges the significance, and in some cases the dominance, of events of greater magnitude and low frequency. Developed for the purposes of palaeontology concerned with sudden and massive extinctions of life forms, a review of the implications for parts of physical geography has been provided by Dury (1980a). Growing interest in storm-generated deposits is evident in sedimentology, and in erosional geomorphology studies of the quantitative effects of events of great magnitude and low frequency have indicated that stream channels and interfluves must be visualized separately, just as there are differences between arid and humid areas. Therefore when considering environmental processes the significance of large events must be considered, the modifications to a strict uniformitarian view must be assimilated and the development of step functions and of catastrophe theory must be attempted.

A process paradigm?

Once the investigation of processes embraces such high-magnitude events it becomes increasingly difficult to distinguish what is timeless and timebound physical geography. On 15 August 1952 a major flood affected the East and West Lyn rivers on the coast of North Devon, had substantial effects on the river channel, the valley and the human settlement, and has been estimated to have a recurrence interval of possibly 1:1000 years (Dobbie and Wolf, 1953). Is such an event appropriate for a process time-scale or more pertinent to a chronology approach? This indeed is one of the implications of investigations of environmental processes because they are now becoming assimilated with new approaches to time and environmental change as outlined in chapter 8 (p. 161). A further consequence arising from studies of process has paradoxically been progress towards a more integrated approach because investigations of one specific process have often proceeded to encompass other processes in the way that soil process investigations have become involved with hillslope processes and with plant–soil moisture relationships and also with boundary layer climatology. It is apparent that even where investigations have commenced within physical geography they have often proceeded to become intertwined with other disciplines. One should not endeavour to place boundaries around physical geography too firmly but an idea of the frontier zone with other disciplines may be essential to shape future developments. Investigations of processes initially functional in nature have begun to remedy deficiencies perceived in the early 1960s, have led to some new visions, and have also led in two directions. On the one hand towards a more realist approach which is essential but cannot go too far because if pursued at other than a meso-scale level is in danger of leading towards natural and biological sciences. On the other hand the path has been from empirical field investigations towards continental or global patterns and this may be one important direction which must continue to be followed because large-scale soil erosion as in China, large-scale deforestation as in the Amazon, atmospheric pollution or global climatic change inspired by changes of carbon dioxide are all contemporary environmental processes on a scale which has not attracted sufficient physical geographers. However that aspect of the current physical geography balance sheet is more appropriate in the next chapter (6).

6

The advent of man

On the one hand the effect of human activity upon earth environment has been very apparent and increasing, and yet on the other hand until the 1950s or 1960s the significance of human activity did not attract much attention by physical geographers, who instead often chose to study environmental change before man, to seek processes largely unmodified by man, or at best to include man as an afterthought or an appendage. This tendency can still be seen in text books where, after a majority of the text devoted to 'straight' physical geography, a final chapter, often brief, may introduce human activity and possibly applied considerations as well! This state of affairs prevailed despite the fact that the theme of the man–environment relation has, according to Harvey (1969), never been far from the heart of geographic research, and for many was the overriding theme.

This paradoxical situation arose notwithstanding great endeavours to retain the strong links between physical and human geography and it was often thought (e.g. Fisher, 1970) that a revival of the regional approach still offered the only method of recementing the two parts of the subject. Such an aspiration prevailed among some geographers at least partly unaware of first, the revival of interest in human activity; secondly, the focus in physical geography research upon the magnitude of human impact; and thirdly, the hazard research that was facilitating closer links between physical and human geography although these links did not impress Johnston (1983c). These three trends combined to form a more environmental physical geography which may have been late but hopefully not too late. In a clear look at man and environment, Hewitt and Hare (1973, pp. 23, 31) assessed the position from their point of view:

> Until quite recently, however, the writers saw this environment in categories that were highly conservative. We accepted the time-honoured division of physical geography into climatology, soil science, geomorphology, and biogeography, holding that these were concerned with the explanatory description of large, tangible and plainly visible parts of the natural landscape, or the directly measurable crude properties of the atmosphere. When, however, we began to teach a course of generalized man–environment relations, in the crisis atmosphere of the early 1970s the physical geography of our background appeared totally inadequate. It simply did not, and does not, offer a framework on which to hang a convincing story . . . the geographer, when he analyses the

material properties of the man–environment systems, must base himself on the central functions of that system, rather than on the traditional divisions of physical geography.

The central functions that Hewitt and Hare identified were partly concerned with human activity and partly involved the systems approach reviewed in the next chapter (p. 140). The albeit late advent of man can be appreciated by outlining the antecedents, the research undertaken to ascertain the magnitude of man, the emphasis upon hazards and upon man-made environments as a basis for a more environmental physical geography.

Antecedents for the investigation of man in physical geography

It is curious that despite frequent allusion to man–environment relations by geographers, they largely chose to ignore the signposts that were evident from the mid nineteenth century onwards and physical geography proceeded largely in isolation from the hand of man. This is very surprising on reflection because there were clear pointers available waiting to be followed. In the introduction to a book on *Man and Environmental Processes* (Gregory and Walling, 1979) the pointers available were resolved into a century of milestones (to 1960), a decade of papers in the 1960s and a decade of readings in the 1970s.

Man and Nature, published in 1864, was undoubtedly the first milestone. Its author, George Perkins Marsh, conceived the book as a

> . . . little volume showing that whereas others think the earth made man, man in fact made the earth. . . . The object of the present volume is: to indicate the character and approximately, the extent of the changes produced by human action in the physical condition of the globe we inhabit; to point out the dangers of imprudence and the necessity of caution in all operations which, on a large scale, interfere with the spontaneous arrangements of the organic or the inorganic world; to suggest the possibility and the importance of the restoration of disturbed harmonies and the material improvement of waste and exhausted regions; and incidentally, to illustrate the doctrine, that man is, in both kind and degree, a power of a higher order than any of the other forms of animated life, which, like him, are nourished at the table of bounteous nature.

To demonstrate the magnitude of changes wrought by man upon nature the book devoted chapters to vegetable and animal species (chapter 2), to woods (3), waters (4), sands (5) and to the projected or possible geographical changes by man (6). This book provided a fundamental basis for the conservation movement (Mumford, 1931), it proved to have a great influence upon the way in which land was visualized and used (Lowenthal, 1965), and its full title *Man and Nature or Physical Geography as Modified by Human Action* clearly indicates the direction in which it was pointing. The clearest reflection of the way in which the full implications of Marsh's book were not appreciated in physical geography is shown in the fact that he is not referred to in relation to the history of the study of landforms before Davis (Chorley, Dunn and Beckinsale, 1964), or in *Explanation in Geography* (Harvey, 1969), in *Geography Its History and Concepts* (Holt-Jensen, 1981) or in *Geography and Geographers: Anglo-American Human Geography* (Johnston, 1979, 1983a). Such omission is completely in sympathy with the trend in physical geography for nearly a century after *Man and Nature*. There were other pointers occasionally, such as Charles Kingsley's book on *Town Geology* published in 1872 and the objection of Kropotkin (1893, p. 350) to the trend to exclude man from physiography.

Man as a Geological Agent, published by R.L. Sherlock in 1922, and a related article (Sherlock, 1923) offered the next major milestone. Sherlock distinguished biological aspects which were not considered in detail in his book, which focused instead upon geological aspects with particular attention devoted to denudation by excavation and attrition, to subsidence, to accumulation, to alterations of the sea coast, to the circulation of water, and to climate and scenery. By focusing upon the impact of man, Sherlock emphasized the contrasts between natural and human denudation and concluded that in a densely populated country such as England (Sherlock, 1922, p. 333):

> Man is many more times more powerful, as an agent of denudation, than all the atmospheric denuding forces combined.

Because of the extent of such powerful human action Sherlock concluded that there were indications that the doctrine of uniformitarianism had been carried too far, believing that the present, which has been so modified by human action cannot readily be the key to the past, when the extent of human modification and human influence was significantly less.

In 1922, the year that Sherlock's book was published, H.H. Barrows gave a presidential address to the Association of American Geographers (Barrows, 1923) entitled *Geography as Human Ecology*. Barrows argued that geography should concentrate on the study of the association between man and environment, it should concentrate on themes which lead towards synthesis with an economic regional geography occupying a central place, but that specialized branches, including geomorphology, climatology and biogeography should be separated from the parent subject. These proposals were not adopted, they were seldom referred to although in the USA much physical geography did not develop in geography departments, but exactly 50 years later a paper with the same title as Barrows's (Chorley, 1973) sought to examine the extent to which the ecological approach to geography provided a unifying link between the human and physical sides of the subject. After reviewing the ecological model and resource management Chorley (1973, p. 167) concluded that the control system (p. 146) could be adopted as an appropriate focus and that this would clearly incorporate human activity and focus upon links between human and physical environment and that:

It is clear, however, that social man is, for better or worse, seizing control of his terrestrial environment and any geographical methodology which does not acknowledge this fact is doomed to in-built obsolescence.

In the 1930s and 1940s, although the impact of man was becoming increasingly more evident from situations like the dust bowl, the Tennessee Valley Authority (TVA), and the much greater use of fertilizers, this situation was not fully acknowledged in physical geography and this is perhaps particularly startling given the appearance of books on soil erosion (e.g. Jacks and Whyte, 1939). However, a major milestone in 1956 was the publication of *Man's Role in Changing the Face of the Earth* (Thomas, 1956) which was in 1193 page book, based upon an interdisciplinary symposium with international participation which was organized by the Wenner-Gren Foundation for Anthropological Research and held at Princeton, NJ in 1955. This monumental achievement acknowledged the work of Marsh and also that of the Russian geographer A.I. Woeikof (1842–1914), who had developed a utilitarian approach to the study of the earth's surface acknowledging the impact of man's activities. The 52 chapters of the work (Thomas, 1956) were organized in three parts: the first retrospective, elaborating the way in which man has changed the face of the earth; the second reviewing the many ways in which processes have been modified; and the third concerned with the prospect raised by the limits on the role of man.

This important volume, subsequently republished in parts, did not perhaps have the immediate effect that might have been expected. However its significance may gradually have been more fully appreciated in 1960s as a series of papers and books provided at least four ingredients which emphasized the necessity to consider man in physical geography. *First* there were a series of general review papers sometimes based on inaugural or presidential addresses and these included a review of *Man and the Natural Environment* (Wilkinson, 1963); advocacy of the need for study of anthropogeomorphology because of the earlier deficiency of studies of the form-creating activity of man and of the influence of man in natural phenomena (Fels, 1965); and a revival of the title used by Sherlock for a review by Jennings (1966) in which he stressed that 'Man as a Geological Agent' is significant because studies of contemporary processes are nearly always heavily biased by anthropogenic effects. Perhaps the most significant paper, also geomorphological, was that in which Brown (1970) characterized man as both a geomorphological process in relation to his direct, purposeful modifications of landforms, and also as indirectly effective through the human influence upon geomorphological processes. Gradually therefore the significance of man began to achieve greater significance in physical geography, although few textbooks before 1970 allocated much space to the effects of human activity. Certainly the notion of the Noosphere (Trusov, 1969), as a new geological epoch initiated as a consequence of man's activity, although capable of wider use in geography (Bird,

1963) did not find unequivocal acceptance.

A *second* strand in research of the 1960s was a consequence of the increasing focus upon processes which gave indications of the magnitude of human activity and also led to the inauguration of research investigations specifically designed to measure the magnitude of man by comparing man-modified and unmodified areas or by measuring one area before, during and after the effects of man. Thus it was possible to investigate processes in a small drainage basin before, during and after the effects of building activity and urbanization (Walling and Gregory, 1970). Equally significant was research on landscape evolution, which was now focused on those time-scales intermediate between those used for earlier chronological studies and those adopted in process investigations. Sometimes this research was located in areas where the effects of human activity had been sustained for many centuries. Thus in the Mediterranean basin the evolution of valleys (Vita-Finzi, 1969) could only be understood by reference to human activity and, more biogeographically, the significance of cultural biogeography was exemplified by the work of Professor D.R. Harris, which was initially in the field of historical ecology but proceeded to embrace the domestication of plants and animals (e.g. Harris, 1968) and which later led to greater contacts with archaeology and to his appointment as Professor of Archaeology in the University of London in 1979.

As a result of the impact of research of this kind a further *third* strand could be detected which devolved upon the investigation of earth hazards and of physical geography from a socioeconomic-viewpoint. Instrumental in launching the study of hazards was Professor Gilbert White's (White *et al.*, 1958) '*Changes in Urban Occupance of Flood Plains in the United States*' which stimulated much later research (p. 132). An excellent illustration of an integrated approach was provided by *Water, Earth and Man* (Chorley, 1969) which was an edited volume reviewing aspects of the physical and the socioeconomic environment that relate to water and which affirmed that (Chorley and Kates, 1969, p. 2):

> .... it is clear without some dialogue between man and the physical environment within a spatial context geography will cease to exist, as a discipline.

To proceed towards such a dialogue Chorley (1969) believed that an integrated body of techniques for physical and human geography and an emphasis upon resources of the physical world such as water could be effected in a volume which (Chorley and Kates, 1969, p. 3) concentrates:

> ... on the physical resource of water in all its spatial and temporal inequalities of occurrence, and by its conceptualization of the many systems subsumed under the hydrological cycle. ... In the development of water as a focus of geographical interest the evolution of a human-oriented physical geography and an environmentally sensitive human geography closely related to resource management is well under way. ... Geographers, freed from the traditional distinction between human and physical geography and with their special

> sensitivity towards water, earth and man, have in these both opportunity and challenge.

In studies of this kind there was evidence of a shift in the attitude to physical environment as environmental perception had a significant influence on the research undertaken by on physical geographers.

A *fourth* and final strand evident in the 1960s was the atmosphere created by the inception of international research programmes and of increasing environmental concern. Programmes dependent upon interdisciplinary as well as international participation included the International Biological Programme (1964–74), the International Hydrological Decade (1965–74) which embraced man's influence as one of its major themes, and the Man and the Biosphere programme. Environmental concern mushroomed during the 1960s and was stimulated by warnings about the impact of human activity and by debates about the extent to which earth resources were finite and therefore the basis for a pessimistic or optimistic future for spaceship earth. Simmons (1978) noted that geographers had taken little notice of the wave of concern for environment which peaked about 1972 and he proceeded to argue in favour of a humanistic biogeography (see p. 124).

At the beginning of the 1970s, in an exploration of the role and relations of physical geography, Chorley (1971) had proposed that control systems offered an approach whereby human activity acts as a regulator in natural systems and this was explored in a major textbook on systems (Chorley and Kennedy, 1971). This book had a substantial influence in physical geography and perhaps initiated a more consolidated attitude to the significance of human activity in physical geography. As the literature available was not entirely adequate for the new acceptance of the significance of human activity, the deficiency was remedied first by a series of collected readings (Coates, 1972, 1973) and Coates (1973, p. 3) noted:

> A twentieth-century innovation has been man's growing awareness of his impact on the environment. With very few exceptions, such as the case of George Perkins Marsh, man before 1900 had a hostile view of the earth; his need was to conquer and subdue nature.

Similarly Detwyler (1971) collected previously published papers to produce *Man's Impact on Environment* and the Association of American Geographers Commission on College Geography conceived a collection of essays devoted to research on environmental problems (Manners and Mikesell, 1974). Subsequently an edited collection reviewed *Man and Environmental Processes* (Gregory and Walling, 1979) and in 1981 A.S. Goudie published *The Human Impact, Man's Role in Environmental Change*. This book (Goudie, 1981b, p. 1):

> . . . . seeks to find out whether, and to what degree, man has during his long tenure of the earth, changed it from its hypothetical pristine condition, for as Yi Fu Tuan (1971) has put it: 'The fact of diminishing nature and of human ubiquity is now obvious'.

The approach adopted by Goudie is to separate physical geography into its conventional divisions of vegetation, animals, soil, water, geomorphology and atmosphere and in a chapter on each to identify and illustrate the effect of human activity on environment and upon processes. In a brief concluding chapter it is shown how study of the human impact can lead to wise management but also that man and nature are difficult to separate, that man is not always responsible for some of the changes with which he is credited, and that environmental impact statements of any kind are difficult to make. This seems something of a disappointing conclusion because some other work has proferred a reorganized approach, in particular either by focusing upon the environment theme or upon the urban environment, which has been covered by *Urbanization and Environment* (Detwyler and Marcus, 1972) by *Urban Geomorphology* (Coates, 1976c) and by *The Urban Environment* (Douglas, 1983).

The foregoing review has been necessary to demonstrate what now seems almost unbelievable; that physical geography for so long contrived to ignore the significance of human activity and thence the potential which associated studies afford. Examples of statements reflecting the recent necessary missionary zeal are included in Table 6.1 Indeed it is indisputable that physical geographers have entered the field later than

Table 6.1: Statements by physical geographers emphasizing the need for the study of human activity

Statement	Source
'It is clear, however, that social man is, for better or worse, seizing control of his terrestrial environment and any geographical methodology which does not acknowledge this fact is doomed to inbuilt obsolescence.'	CHORLEY (1973b)
'Only a relatively small part of this literature has stressed man's role as a geomorphic agent, exerting a notable effect on the earth's crust. On the whole, specific attention to this geomorphic rather than ecologic aspect has been neglected. . . . The study of man's geomorphic activities, . . . is one of the paradoxical "recognized yet unrecognized subjects".'	GOLOMB and EDER (1964)
'If geography is to play a full role in the study of environmental problems it is important that greater stress be given to the biosphere, which constitutes a vital resource base for man, and which has been altered more extensively by human activities than most other elements of the environment.'	HILL, 1975
'. . . a more meaningful and relevant physical geography may emerge as the product of a new generation of physical geographers who are willing and able to face up to the contemporary needs of the whole subject, and who are prepared to concentrate on the areas of physical reality which are especially relevant to the man-oriented geography. It is in the extinction of the traditional division between physical and human geography that new types of collaborative synthesis can arise.'	CHORLEY and KATES (1969)

they should have done. Their contribution so far has emphasized reviews of human impacts often reflecting research in other disciplines rather than appropriate fundamental impact studies by physical geographers. This indeed is one of the reasons why Hare in 1969 noted 'sometimes I think that Geography as a science deliberately stays out of phase with the climate of the times.'

The magnitude of man

In one of the most recent textbooks (Goudie, 1981b) the approach adopted was to review the impact of man on environment and processes generally but in this section the purpose is much narrower: to indicate some of the research investigations pursued by physical geographers and to refer to the attitudes which this engendered.

Biogeography provides the most obvious area in which the significance of human activity should be particularly demanding of study and there have been two particular themes, albeit increasingly related, which have proceeded in this direction. *First* were studies of Quaternary ecology dedicated to reconstruction of the sequence of vegetation systems in particular areas and often using palynology as a technique of central significance. Physical geographers together with biologists, limnologists and other researchers were able to establish the nature of the flora and fauna in the late Quaternary and at the transition during the early Flandrian from tundra to closed deciduous woodland. This research successfully established the major stages of vegetational change but also necessarily included the influence of human activity, which was effective through modification of the biogeographic systems. This naturally fostered closer links with archaeology (Simmons and Tooley, 1981). In Britain the impact of prehistoric man has been considered by physical geographers and in a thematic issue of the *Transactions of the Institute of British Geographers* devoted to this research (Curtis and Simmons, 1976, p. 257) it was explained that:

> The papers presented . . . belong mostly to the tradition of elucidating the effects of early man upon his surrounding landscapes. This might seem a particularly appropriate task for geographers although similar work has been the province of biologists, archaeologists and geologists.

Work of this kind demonstrated the significance of the Bronze and Iron Ages in reducing the forest cover of the British landscape particularly in upland Britain, where blanket peat growth may have been influenced by deforestation rather than simply by climatic change (Simmons, 1980). Although initially research was often focused upon a single pollen site, the subsequent research was able to proceed towards the regional assessment of past vegetation not only in Britain but also in overseas areas such as the tropical rain forest (Flenley, 1979). In this research, palynology was dominant for many years but subsequently enormous potential has been offered by research developed by F. Oldfield on mineral magnetic properties which are valuable because many magnetic properties are environmentally diagnostic, are preserved for long periods in many situations, and have parameters which are easy to measure. Initially applied in the context of lakes and their drainage basins which can be used as units of sediment-based ecological study (Oldfield, 1977), the techniques have now been shown to have widespread application to sediment correlation in lake sediments, to differentiation of weathering and pedogenesis, to the identification of sediment sources, and there may subsequently be application of mineral magnetic measurements to long sediment sequences from major lake basins in non-glaciated regions, to near-shore marine sediments in morphogenetically dynamic areas, and also to cave sediments, alluvial fills, river terrace sequences and to loess successions. Oldfield (1983a, p. 150) also indicates:

> The ubiquity and sensitivity of magnetic minerals, the speed and versatility of measuring equipment and the persistence of magnetic linkages between source and sediment make the emerging methodology ideally suited to both process- and reconstruction-oriented catchment studies and more especially to that integration of the two approaches so strongly advocated in recent time.

Experience of these techniques has recently prompted Oldfield (1983b) to propose a steady-state model of ecosystem change related to man's impact on environment as an additional alternative to more familiar successional and cyclic models, and this is more appropriately considered in relation to other time-bound developments (chapter 8, p. 182).

A *second* theme evident in biogeography has been the progress made towards cultural biogeography and historical ecology. This approach has been very evident in the United States, in the most recently settled areas, and has involved reconstruction of changes in biogeographical systems in recent decades and centuries incorporating the influence of human activity. Similarly Murton (1968) reconstructed the immediate pre-European vegetation on the east coast of the North Island of New Zealand. However, particularly in long-settled countries, it is difficult to differentiate such biogeographical studies from those by historical geographers or economic historians and it was a historical geographer (Darby, 1956) who mapped and compared the distribution of forest in AD 900 with that in AD 1900. However, progress in research in this second theme also infuenced the presentation of biogeography was reflected in textbooks. Although a number sought to focus upon the original vegetation, and therefore to relegate human activity to a chapter or afterthought, this was becoming increasingly invalid as the effects of human activity were so substantial. I.G. Simmons in his book devised a title *Biogeography: Natural and Cultural* which encapsulates the approach that he promoted (Simmons, 1979a). Having suggested that within geography, biogeography teaching has lacked a crystallizing focus, he contended that (Simmons, 1979a, p. 1).

> Instead there has been something of a supermarket approach: according to the interests of the teacher a number of packages have been bought off the shelves and put in the course trolley. The major world soil–climate vegetation units; soils, the 'natural' vegetation units of a region or country; some vegetation history; and the consideration of energy and chemical element flows through ecological systems, have been put together and presented as 'biogeography' at the check-out point of the examination paper.

Simmons proceeds to offer a stronger conceptual framework. Acknowledging the concepts of Dansereau (1957) that man creates new genotypes and new ecosystems, he considers nature without man as natural biogeography in Part I of his book and then uses this as a datum against which to set the larger Part II in which cultural biogeography is concerned with the effects of man in changing the genetic make-up of plants and animals, in redistributing them over the earth's surface, and in altering the structure of many ecosystems. Although Simmons indicates that such an approach may encompass only part of the traditions of study in geography, it would allow man's impact on individual taxa and ecosystems, on biotic resources their conservation and protection (Simmons, 1980, p. 148)

> to find a place in a coherent whole, which relates to the concerns of the world about us. Nevertheless, I expect that the methodological pluralism which has characterized our biogeography during my time will continue, and diversity of both material and outlook must in any final reckoning be a source of strength.

There is a danger in indulging in too great a diversity of material and outlook and, as biogeographers have been few in number, diversity could lead to the continuing omission of the study of the impact of human activity. However Simmons's (1979a) approach is one which leads productively towards consideration of natural resources (Simmons, 1974) and towards involvement in conservation (Simmons, 1979b) as reviewed in a later chapter (p. 186).

In the study of soil geography two strands similar to those in biogeography may be discerned: one arising from studies of soil profiles and their evolution and one more concerned with recent historical changes. Associated with late Quaternary vegetation changes were changes of pedogenesis and of soil systems. In Britain, for example, podzolization succeeded brown earth formation, and because human action was involved in the vegetation change there was also an effect upon soil profiles (Bridges, 1978b). Recent historical changes of soils are more numerous than the degree of research attention accorded by physical geographers might suggest but the effects of salinity, drainage, fertilizers, pollution and land use change all have considerable significance for soil pedons and soil processes. Bearing in mind the magnitude of soil erosion which has occurred in sensitive areas including the Dust Bowl, parts of China and of Africa it is surprising that more attention has not been directed to this aspect of the human impact. It has been argued that special attention should be focused upon the resilience and potential for recovery of the soil profile in view of the inputs induced by man (Trudgill, 1977, chapter 8), and the importance of the problem is underlined by Toy (1982) in a review of accelerated erosion when he concludes that such erosion can be considered to be the pre-eminent environmental problem in the United States by virtue of its widespread occurrence and cumulative cost. A review by one geographer who has contributed to the development and application of indices of soil erodibility (Bryan, 1978) includes the aphorism (Bryan, 1979, p. 207):

> It has been said that if each soil conservationist stopped the movement of one grain of soil for each word he has written on the topic, the problem of soil erosion would disappear. It is true that despite close attention from numerous scientists around the world for the past 60 years, soil erosion is now perhaps more widespread than ever before.

Despite the numbers of scientists involved, physical geographers have not perhaps been as prominent as they should have been in the investigation of the human impact upon soil systems.

Whereas the long-term significance of human activity upon plants and animals and soil characteristics and distributions has provided one research focus, in climatology the emphasis has perhaps been more evident on a spatial scale with the impact of human activity first appreciated at the local scale but then subsequently extended to the meso and thence to the world scale. With this progression the atmosphere has increasingly been viewed as a resource (Chandler, 1970) and research interests have concentrated upon (Chandler, 1970, p. 3):

> . . . those aspects of the atmospheric sciences, which, to my mind, geographers with a knowledge of physics and mathematics are especially well equipped to study, namely the particular conditions of the planetary boundary layer. . . . In draining the marsh, clearing the wood, cultivating the fields, and flooding the valleys, man has inadvertently changed the thermal, hydrological and roughness parameters of the earth's surface and the chemical composition of the air. . . . There are indeed very few 'natural' boundary layer climates remaining. . . .

Although some recent textbooks have included human activity in a final concluding chapter (e.g. Lockwood, 1979a) others have accorded the significance of human activity a more central position. Thus Oke (1978) emphasized the physical basis of physical, topo-, local-, meso- and regional climates in the boundary layer climates of the lowest kilometre of the atmosphere, then proceeded to deal with natural climates of non-vegetated and vegetated surfaces, topographically affected climates and the climates of animals, and then progressed to man-modified environments embracing conscious and unconscious modification of boundary-layer climates particularly those including urban climates. Subjects to which physical geographers have contributed include changes in the radiation balance, changes in precipitation including conscious and inadvertent ones, man-made climates and pollution. Perhaps in the area of man-made climates has the physical geographer climatologist contribution been most significant and the work of T.J. Chandler (1965) on the *Climate of London* stands as an exemplary model in this field, and subsequent research by B.W. Atkinson on thunderstorm activity (see Atkinson, 1979) clearly demonstrates the kind of contribution that can be made to document the inadvertent effects of man. Until recently the contribution by physical geographers to the investigation of atmospheric pollution by carbon dioxide, aerosols, fluorcarbons, thermal pollution, and other pollutants, and the effects which they have had, has been less evident. However as Chandler, who was a member of the Royal Commission on Environmental Pollution established in the UK in 1970 (Chandler, 1976), has remarked (Chandler, 1970, p. 10):

> . . . . atmospheric pollution has been with us since the creation; it became much worse . . . with the Garden of Eden and infinitely worse following the mechanical ingenuity of James Watt.

Considerable interest is now centred upon these problems and it is appropriate that physical geographers should be very aware of the work being undertaken and the work remaining. In 1980 an international meeting in Villach, Austria was called by the International Council of Scientific Unions (ICSU), the United Nations Environment Programme (UNEP) and the World Meteorological Organization (WMO) to prepare an assessment of the impact of CO_2 on global climate. They concluded that because approximately half of the 5 gigatons of carbon released into the atmosphere each year by the consumption of fossil fuel remains in the atmosphere, the world temperature has already been affected. More specifically they suggested that the impact of change will vary differentially over the earth's surface, and that further research is needed on likely consumption of fossil fuel in the next century, on prospective modes of management of the global biosphere over the next century, on quantitative clarification of the carbon cycle, on climatic response to increasing CO_2 in the atmosphere, and on the potential impact of climatic change (Nature and Resources, 1981). It was believed that a major international interdisciplinary research effort was needed and in such areas physical geographers should surely be poised to participate.

In geomorphology assessments of human impact have really been focused upon four themes if one includes consideration of hydrology and hydrogeomorphology at the same time. First has been research focused upon chronology of landscape change which has (Butzer, 1974; Vita-Finzi, 1969) demonstrated how human activity was a pertinent factor influencing the course of alluvial chronology and in Greece Davidson (1980a) has demonstrated that erosion was underway during the Bronze Age but that spatial variations were substantial. In more recent periods of decades there have been, secondly, many investigations of the impact of human activity for example upon river channels (Gregory, 1981; Park, 1981) and of changes during decades rather than millenia. Many of these are mentioned in chapter 8 (p. 173) but a particularly important viewpoint offered by

Strahler in 1956, provided fundamental guidelines which acknowledged earlier work by engineers, hydrologists, soil scientists and geologists and proceeded to formulate an improved understanding of fluvially produced landscapes. This included a diagram of drainage basin transformation from low to high density which related the severe gullying that could occur on slopes to the aggradation along the main valley axis.

Thirdly there have been studies of geomorphological processes affected by human activity and particularly important stimuli were produced by Wolman (1967a) who suggested the way in which sediment yield varied at the present time between urban and non-urban areas and extended this to provide a model of change of sediment yield in the northeast of the USA since 1700, and also by Douglas (1969). In his review of man, vegetation and the sediment yields of rivers Douglas suggested that little account had previously been taken of human influence and his studies of rates of erosion in a wide range of climatic conditions in eastern Australia suggested that the present sediment yields are far in excess of those which may have prevailed in the geological past. Methods of calculation of such denudation rates have been greatly refined by Walling (e.g. Walling and Webb, 1983) who has also reviewed the ways in which hydrological systems are modified by man, and results from catchment experiments have documented the effects of land clearance, land drainage, recreational pressure, strip mining, conservation measures and urbanization (Walling, 1979). In the easten part of the USA the significance of conservation measures has been demonstrated at two different scales (Fig. 6.1). The magnitude of human activity has also been reviewed with respect to coasts where sea walls, breakwaters, dredging and dumping, sediment supply, and vegetation effects can all be significantly affected by human activity (Bird, 1979) and the beach sediment budget can be profoundly changed (Clayton, 1980b). Whereas effects upon coastal processes have been long considered at least in qualitative terms, the significance of human activity in affecting permafrost as illustrated in Canada (Brown, 1970); and in influencing endogenetic processes (Coates, 1980), by loading effects from dams and reservoirs, water injections, irrigation and structures, and by withdrawal effects due to groundwater, oil and gas, and mineral extraction, and by surface excavation effects have been appreciated more recently. Important inputs to environmental management are afforded by studies of human influence upon permafrost and upon endogenetic processes.

A fourth theme in geomorphology has been concerned to demonstrate the world impact of human activity. Although no geomorphology texts have been structured to emphasize man, as in the case of Simmons (1979a) in biogeography and Oke (1978) in climatology, some progress towards global assessment is being made. Thus Brown (1970) estimated world-wide natural erosion rates to range from 12 to 1500 $m^3km^{-2}yr^{-1}$ and man-induced rates to range from 1500 to 85,000 $m^3km^{-2}yr^{-1}$ and the results from a wide variety of areas were reviewed by Gregory and Walling (1973,

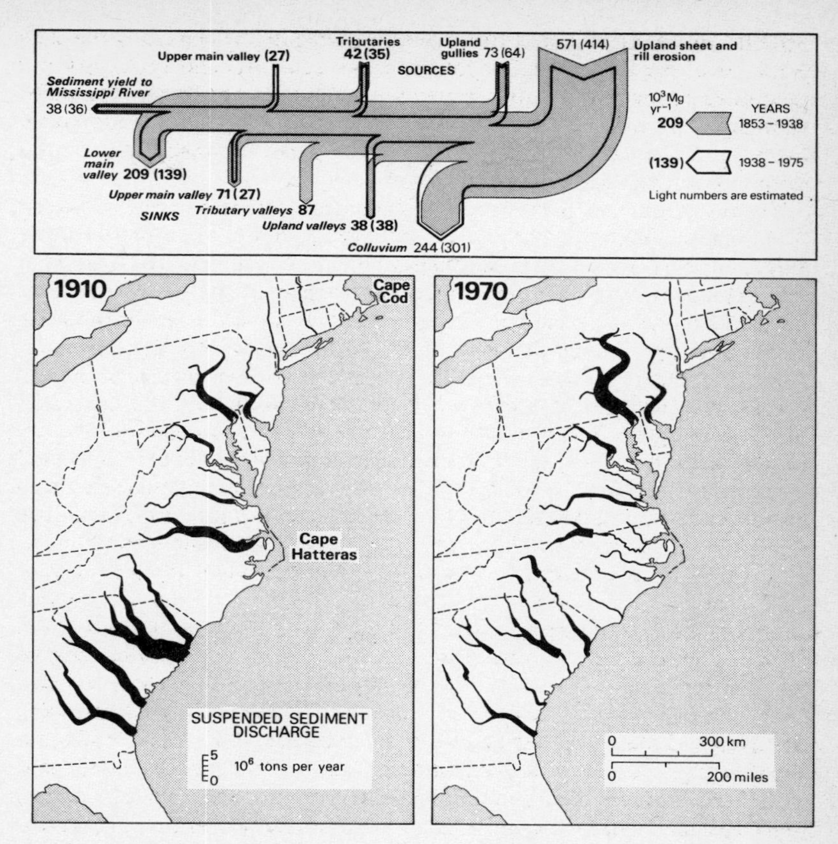

Figure 6.1: The influence of human activity upon sediment transfer. In the cartogram the magnitude of sediment production from different sources in Cook Creek is compared before and after conservation measures (after Trimble, 1983). In the lower two maps suspended sediment discharge is reduced by 1970 after the implementation of conservation measures and after changes in land use (after Meade and Trimble, 1974)

pp. 342–58). Demek (1973) concluded that the effects of human society on the development of the earth's relief already exceeds the effects of natural geomorphological processes and that, of the dryland surface of the earth, 55 per cent is intensively used by man, 30 per cent is partly modified by man, and the remaining 15 per cent is either unmodified or slightly modified. Perhaps recognition of a field of urban physical geography is an indication of the way in which geomorphologists are beginning to perceive a world pattern.

When surveying the focus of attention of physical geographers it is evident that the major part of the earth's surface, the oceans, has been ignored in physical geography research. Although a textbook on oceanography was produced by a physical geographer and was subsequently divided into two volumes, (King, 1962; 1975) the separate discipline of oceanography has been responsible in physical oceanography for research on sea level and its changes; tides, waves, currents, the physical and chemical properties of sea water as well as the form of the sea floor and marine sediments, and in biological oceanography for research on ocean flora and fauna. Physical geographers have also tended, with some exceptions in countries where there are a large number of lakes, to leave the study of lakes to the limnologists. However, rather more attention has been devoted to what R.U. Cooke characterizes as an 'empty quarter' (Cooke, 1976). The deserts of the world which comprise this empty quarter demonstrate a sensitivity to human impact but one which is not easily separated from the effects of fluctuations in climate, as studies of arroyo development have clearly shown (Cooke and Reeves, 1976). Much recent concern has been focused upon desertification, which is taken to mean (Grove, 1977, p. 299):

> the spread of desert conditions for whatever reason, desert being land with sparse vegetation and very low productivity associated with aridity, the degradation being persistent or in extreme cases irreversible.

Desertification arises from climatic change and from unwise use of land by man and has been the subject of a number of international and interdisciplinary meetings. At the International Geographical Union meeting in Montreal in 1972 the Arid Zone Commission was succeeded by a working group established to study desertification in and around arid lands with J.A. Mabbut as chairman and the aim was to 'collect evidence on the nature and causes of environmental changes constituting extension of deserts into marginal regions or an intensification of desert conditions within desert regions' (Mabbut, 1976). A number of UNESCO programmes have embraced this research area and the MAB (Man and the Biosphere) Programme has brought attention to the impact of human activities and land use practises on arid and semi-arid regions.

Earth hazards

Desertification is of course, at least in part, an instance of drought as an earth hazard. As studies of human impact and reviews of the magnitude of this impact encouraged physical geography to move towards applied problems it was necessary to alter attitudes to physical environment and this occurred as a result of three tendencies. First there was the trend to focus upon extreme events because these are the ones that may occasion damage and costs, they are the ones for which landscape management strategies may be designed, and this was a departure from the earlier emphasis upon the average experience. In addition to the immediately obvious atmospheric hazards it was possible to include less immediately obvious ones associated with volcanoes (Clapperton, 1972) and earthquakes. Whereas endogenetic processes had not been studied as extensively as exogenetic ones this now gave a focus for study by physical geographers as illustrated by the analysis of eruptions by Etna (Clapperton, 1972). The study of volcanoes in space and time has been advocated by Clapperton (1972) to build a bridge between volcanology and volcanic geomorphology which would be analogous to these fields developed in glacial and in fluvial geomorphology. Secondly a trend reflected in the juxtaposition of investigations of physical environment and of the socio-economic relevance was exemplified by *Water, Earth and Man* (Chorley, 1969) and by the *Value of the Weather* in which Maunder (1970) introduced the range of atmospheric hazards and then proceeded to evaluate their cost in terms of studies of economic impact and the costs of hazard relief. Maunder expressed his objective as (Maunder, 1970, pp. xxi, 2).

> The purpose of this book is to bring together, for the first time, the most significant and pertinent association between man's economic and social activities, and the variation in his atmospheric environment . . . specific attention is focused on economic activities and the weather, the economic analysis of weather, and the benefits and costs of weather information, including weather forecasting and weather modification. In addition some of the many sociological, physiological, political planning and legal aspects of man's atmospheric resources are discussed. . . . it is hoped to provide a starting point for econoclimatic studies on a national and international scale, and a greater appreciation of the value of the atmospheric environment.

A later study of the hurricane impact (Simpson and Riehl, 1981) also epitomizes the way in which a focus can be placed on the nature of the hazard and the socioeconomic cost.

A third trend was centred upon growing awareness of the difference between the real world and the way in which environment was perceived because it was such perception which influenced decision-making and therefore management. Following from this it was therefore axiomatic that knowledge of environment was time-dependent and the environmental perception at the time of decision-taking could be very significant. Although early progress in study of environmental perception often related to socioeconomic geography such as the attitudes of farmers to the drought hazard on the Great Plains (Saarinen, 1966), later research concentrated more upon the physical environment and the fascinating evaluation of myth and reality in the context of a volcanic eruption in Papua New Guinea (Blong, 1982) has already been referred to.

These three ingredients collectively formed the background in which natural hazards research crystallized, although as noted earlier the origins clearly arose in North America and particularly derived from the work of

Gilbert White at the University of Chicago, later at the University of Colorado, and leader of the Commission on Man and Environment (1968–) of the International Geographical Union. Professor White's earlier research was directed to flood plains and following the Flood Control Act of 1936 the research group asked the following questions (White, 1973, p. 197):

1. What is the nature of the physical hazard involved in extreme fluctuations in streamflow?
2. What types of adjustments has man made to those fluctuations?
3. What is the total range of possible adjustments which man theoretically could make to those fluctuations?
4. What accounts for the difference in adoption of adjustments from place to place and time to time?
5. What would be the effect of changing the public policy insofar as it constitutes a social guide to the conditions in which individuals or groups chose from the possible adjustments?

The study classified adjustments as modifying the cause, modifying the loss or distributing the loss and established that while flood control expenditures had multiplied the level of flood damages had risen, and that the natural purpose of reducing the toll of flood losses by building flood control projects had not been achieved (White, 1973). Subsequent research on a spectrum of natural hazards was collected during work of the IGU Commission (White, 1974) in which natural hazard was defined (White, 1974, p. 4) as:

> . . . . an interaction of people and nature governed by the coexistent state of adjustment in the human use system and the state of nature in the natural events system. Extreme events which exceed the normal capacity of the human system to reflect, absorb or buffer them are inherent in hazard. An *extreme event* was taken to be any event in a geophysical system displaying relatively high variance from the mean.

Although White (1973) noted that in the 1960s geographers turned away from some environmental problems just as specialists in neighbouring fields discovered these issues, nevertheless research on hazards did demonstrate that (White, 1973, p. 213):

> . . . . if environmental problems are pursued rigorously enough and with sufficient attention to likely contributions from other disciplines they may foster constructive alterations in public policy but at the same time may stimulate new research and refinement of research methodology to the benefit of geographic discipline.

Subsequently a number of books have summarized this and other work on earth hazards (e.g. Whittow, 1980; Perry, 1981) but two singular ones are worthy of special attention. Hewitt and Burton (1971) analysed the record for southwestern Ontario and found that in a 50-year period there would be 1 severe drought, 2 major windstorms, 5 severe snowstorms, 8 severe hurricanes, 10 severe glaze storms, 16 severe floods, 25 severe hailstorms and 39 tornadoes. They therefore defined the hazardousness of a place as the complex of conditions which define the hazardous part of a region's environment. Hazards were taken to be simple, which include a single damaging element such as wind, rain, floodwater or earth tremor; compound, which involves several elements acting together above their respective damage thresholds such as the wind, hail and lightning of a severe storm; and multiple, when elements of different kinds coincide accidentally or follow one another as a hurricane may be succeeded by landslides and floods. A further progressive step was made by Burton, Kates and White (1978) in *The Environment as Hazard*. Beginning from a consideration which suggested that the natural environment is becoming more hazardous in a number of complex ways because losses are rising, catastrophe potential is enlarging, and the cost falls inequitably amongst the nations of the world, they proceeded from hazard experience to consideration of choice on an individual, collective, national and international level. Although they concluded (Burton, Kates and White, 1978, p. 221) that an increase in disaster may continue in the decade of the 1980s and that reduction of disaster potential cannot easily be achieved, although loss of life will be reduced substantially, loss of property is most likely to occur in rapidly developing societies and overall (Burton, Kates and White, 1978, p. 223):

> The forces propelling the world toward more and greater disasters will continue to outweigh by a wide margin the forces promoting a wise choice of adjustments to hazard. There is hope for a safer environment, but it cannot be achieved easily or soon.

Natural hazard research necessarily focuses upon the interrelation of geophysical events and human activity and as such is an important feature of recent research and one that some such as Parker and Harding (1979) have proclaimed to be of central and traditional concern to geographers and their view is included in Table 6.2.

Table 6.2: Views on the importance of natural hazards for geography

'Natural hazard studies provide a novel and imaginative vehicle for teaching aspects of physical and human geography. They draw attention to the dynamic relationship between man and environment, provide an opportunity to introduce the concept of perception and enhance general environmental awareness which is an important element of environmental education.'	D.J. PARKER and D.M. HARDING (1979)
'An important task . . . is to establish the place of damaging events within the overall context of man's ecology. Many of the natural elements which cause damage are, under more normal circumstances, sources of livelihood and man's activities are closely integrated with them.'	K. HEWITT and I. BURTON (1971)

The outcome of research on natural hazards has been the emergence of a dominant view according to Hewitt (1983) and this convergence of thinking conforms to the idea of a paradigm. This dominant view is that disaster itself is attributed to nature and embraces three main areas:

1 an unprecedented commitment to the monitoring and scientific understanding of geophysical processes – geologic, hydrologic and atmospheric – as the foundation for dealing with their human significance and impacts. Here the most immediate goal in relation to hazards is that of prediction,
2 planning and managerial activities to contain the geophysical processes where possible,
3 emergency measures, involving disaster plans and the establishment of organization for relief and rehabilitation.

Although admittedly criticizing some of his own work Hewitt (1983) believes that the perspective of this dominant view could be the single greatest impediment to improvement of the quality and effectiveness of natural disasters research, because it fails to recognize how the roots and occurrence of contemporary disasters depend upon the way 'normal everyday life turns out to have become abnormal'. In a series of chapters edited by Hewitt (1983) it is demonstrated how natural hazard is not uniquely dependent upon geophysical processes, human awareness is not dependent upon the geophysical conditions, and reaction to disaster may be dependent upon ongoing social order rather than explained by conditions or behaviour to calamitous events. Hewitt (1983) is then led to conclude that most disasters are characteristic rather than accidental features of places and societies where they occur; risk arises from ordinary life rather than rareness, and natural extremes are more to be expected than many of the social developments that pervade everyday life. This provides an alternative perspective but still one in which man and hazards interact, which is underlined by part of the quotation used at the beginning of Hewitt's (1983) volume:

> It's because people know so little about themselves that their knowledge of nature is so little use to them. Bertolt Brecht (1965) *The Messing Kauf Dialogues* (London: Methuen)

Urban environments

Although texts devoted to the human impact (e.g. Goudie, 1981b; Gregory and Walling 1979) and those devoted to branches of physical geography have not yet emphasized urban physical geography, claims have been made by some writers that the urban environment is sufficiently distinctive to warrant attention as a specific milieu. The books referred to earlier by Detwyler and Marcus (1972) and by Coates (1976a) together with a volume of readings on urban regions (Coates, 1973) constitute a considerable directive inclined towards an urban physical geography. This was reinforced by research investigations by physical geographers on urban climates, and the seminal work by Chandler (1965) on *The Climate of London* stimulated other physical geographers to document the magnitude and character of urban heat islands, of precipitation modification, of atmospheric pollution and of air movement. Studies have been conscious of the work in other disciplines such as the Metropolitan Meteorological Experiment (METROMEX) which is based in St Louis (Chagnon, Huff, Schickedanz and Vogel, 1977) and has greatly increased knowledge of atmospheric processes within urban areas and investigations have also led towards applications of knowledge gained from the way in which a city generates its own climate so that Chandler (1976) in a review of urban climatology in relation to urban design suggested:

> The reason for the neglect of climatic consideration has been partly the relative youthfulness of the science of urban climatology, and partly the relatively weak communication links that presently exist between climatology and planning. But faced with the exponential growth of the world's population and the accelerating pace of urbanization, it is clear that our cities must, where appropriate, be purposefully planned in order to optimize the environment of urban areas and avoid a series of structural and functional design failures. Climate is an essential element in this planning.

Also in 1976 a symposium on physical problems of the urban environment (Chandler, Cooke and Douglas, 1976, p. 57) concluded that:

> The conversion of land to urban uses involves considerable modification of the natural environmental system, particularly with respect to its geological and geomorphological bases, its hydrological characteristics and the nature of the boundary layer of the atmosphere. . . . Physical geography can usefully contribute to the determination of public policies, with respect to the management and development of urbanized areas.

In the area of hydrology have been some of the most specific contributions made by physical geographers including the analysis of increased discharges (Hollis, 1975; Walling and Gregory, 1970; Walling 1979a), investigation of sediment yield (Walling, 1974), research extending into the area of water quality and pollutants related to urban source areas (Ellis, 1979) and analysis of river channel changes downstream from urban areas (Leopold 1973; Gregory, 1981). Indeed in an edited volume devoted to *Man's Impact on the Hydrological Cycle in the United Kingdom* Hollis (1979) collected results of current scientific investigation for the benefit of scientists and managers and separated the chapters into sections dealing with rural and urban environments respectively, a subdivision which was also used in the hydrological field (IAHS, 1974). In the case of urban climatology and urban hydrology there is obviously an important research contribution to be made by physical geographers and this is made by undertaking work in knowledge of, or close association with, scientists in other disciplines.

However one geographer, Douglas (1981, p. 315), has argued that:

> Understanding of the dynamics of the biophysical components of the city and the way their functioning impinges on people is a vital part of urban studies.

In his attempt to determine whether the methods and concepts of ecology and physical geography can contribute effectively to analysis of urban issues in the context of the city as an integrated open system of living things interacting with their physical environment, Douglas (1981) argues that an attempt must be made to link the city as a habitat or an ecosystem with the city as a social system and he develops an ecosystem approach to the study of cities. This approach, which had earlier been used in relation to other fields such as architecture (Knowles, 1974) embraces population ecology, system ecology, the city as a habitat, and then energy and material transfer in cities are used to highlight spatial contrasts within the city and between cities and also to differentiate cities from rural environments. Douglas quotes support for study of urban areas such as that by Bunge (1973):

> Physical geography is much needed in an urban setting. . . . Cities are a karst topography with sewers performing precisely the function of limestone caves in Yugoslavia, which causes a parched physical environment, especially in city centres,

but also notes that some such as Young (1974) have argued that geographers have failed to realize the potential of contributing to human ecology because they were overconcerned with environmental determinism. Subsequently Douglas has developed these ideas in a book which collects together much material in the urban environment from a physical geography point of view but perhaps also leads towards his view that (Douglas, 1981, p. 360):

> Perhaps more than in other environments the analysis of city problems reveals the inadequacy of attempts at understanding through blinkered, disciplinary approaches.

In this book Douglas (1983) elaborates an ecosystem view of the city. He uses the background of the economic system as an ecosystem and the city as a dependent system, to proceed to the energy balance, water balance, mass balance geomorphology, biogeography and waste disposal of the city prior to looking at geographical aspects of urban health and disease and at management and planning designed to reduce environmental hazards. Douglas (1983, p. 202) tends to the view that:

> in academic scholarship the divorce between the 'two cultures', the humanities and the sciences, sadly persists, and only at the practitioner level is there some collaboration between the social worker and the public health engineer. . . . In examining cities one cannot be simply scientific or simply sociological. . . . To learn about or to teach about cities without considering both the biophysical environment and the social environment is downright unscholarly.

However the compartmentalized approach, rather reminiscent of that followed by Goudie (1981b), may not be as integrative as Douglas would hope, although the use of a systems paradigm and energy flows as discussed in chapter 7 (p. 153) may provide a more integrative solution.

The physical geography of the city in general can now complement studies already undertaken of specific cities. Los Angeles has been particularly attractive as a case study and has been outlined by Cooke (1977) and used by Whittow (1980) as an environment which experiences an accumulation of hazards. A focus upon the urban physical environment could be seen as a replacement for regional geography courses of the past but as one reviewer has noted (Clayton, 1984, p. 26):

> The urban environment is not defined as readily as a region such as North America or Australia but it is a way of studying part of the world that has some unifying attribute. Whether the result is really much better than our much maligned regional courses is hard to say. It seems to remain unduly descriptive and obsessed with detailed case studies.

Environmental physical geography

Until the advent of man in physical geography it could be argued that physical geographers had escaped from the effects of human activity by concentrating their endeavours upon rural and unmodified spatial areas and upon time scales prior to the time when human activity began to exercise a significant influence. Some physical geographers have envisaged that dangers lie not only in this refuge in the unmodified environment but also in the fact that research investigations in modified situations may involve small-scale studies of, for example, gully pots in urban stormwater drainage systems. One answer to such a detailed approach has been to revitalize the notion of large-scale geomorphology for example (Gardner and Scoging, 1983) but another has been to concentrate upon human activity as providing a focus centred around the central system (p. 156). Even in this regard however it could be argued that physical geography has not reacted sufficiently to undertake research on major global problems such as the increase of atmospheric CO_2, the incidence of acid rain, the demise of the Amazon rainforest, the implications of world soil erosion, or the general field of environmental pollution. It was noted earlier that Hare in 1969 had noted how geography often endeavours to stay out of phase with the climate of the times and in a later paper (Hare, 1980) he directed attention to whether the planetary environment was fragile or sturdy. He concluded that the planet's habitability is under attack but its natural resilience is probably greater than is normally assumed, and that in the environmental field geographers have made contributions in environmental perception and hazard assessment and also in the area of unified physical and biogeographic research:

> One has to admit, nevertheless, that geographers as a group have not taken the lead in the interdisciplinary study of the human environment. More often it has been the engineers and the ecologists who have done so with lawyers, economists and political scientists sometimes joining in (Hare, 1980, p. 381).

Referring to widespread stresses Hare reviewed the carrying capacity of the earth, and planetary air pollution and climatic change with considerable vision and in ending on a fairly optimistic note commented:

> For the geographer, the opportunity to analyse and predict this resilience is a marvellous challenge. Humanity and nature exist in constant interaction. We have understood this crucial point much longer than most of our colleagues, who tend to think in terms of the impact of nature on society, or the reverse. In the past decade it has seemed to me, perhaps over-optimistically, that the logic of interaction was taking command over what we teach, and the things that we try to do. The sense that our field was disintegrating had receded. It has a unity rooted in natural logic.

A more integrated approach to the environment has also been preferred by Coates who during the 1970s, which he characterized as the Environmental Decade, edited a volume (Coates, 1971) on *Environmental Geomorphology*, which was defined as:

> .... the practical use of geomorphology for the solution of problems where man wishes to transform landforms or to use and change surficial processes. ... In addition environmental geomorphology includes extraction of surficial materials and protection of certain landscapes, such as beaches, which benefit man. The goal for geomorphic environmental studies is to minimize topographic distortions and to understand the interrelated processes necessary in restoration, or maintenance, of the natural balance.

This volume, and volumes of readings on *Environmental Geomorphology and Landscape Conservation* (Coates, 1972, 1973), have revitalized the scope and focus of geomorphology and the approach contained could be applied to physical geography as a whole. However such an environmental physical geography is more a focus than a means. In the urban environment (Douglas, 1983) and in an ecosystem approach to the city (Douglas, 1981) Douglas employs an ecosystem or systems approach and that, particularly the control system, may provide the means for study in physical geography and so that is appropriately the subject for chapter 7. When looking at geographic geomorphology in the 1980s Graf, Trimble, Toy and Costa, (1980) observed that the human factor in geomorphology received insufficient attention until the late 1960s but that researchers are now recognizing the effect of human activity in many processes ranging from relatively minor disturbances to almost complete control. They concluded that, especially in relation to fluvial processes, explanations that ignore the role of human activities run the risk of eliminating one of the most significant variables and perhaps an appropriate note on which to end this chapter is their affirmation that (Graf, Trimble, Toy and Costa, 1980, p. 281):

We believe that no other group engaged in geomorphic research is as well qualified to grapple with the human factor as are geographers. Our broad training in both physical and cultural systems and our appreciation of landscape change in the natural and human senses give us perspectives and insight that are rarely found in other disciplines. Geographers need to work closely with engineers and geologists in order to share with them such wider ranging concepts as spatial analysis and emphasis on the man–land interface. These concepts are endemic to geography, but they may be quite foreign to other workers.

7

The environmental system - All systems go?

The themes of quantitative methods, of chronology, of processes and of human activity as reviewed in the last four chapters all have amongst their adherents some who would claim that one of these four was the dominant paradigm for physical geographers. However, each of those themes requires a unifying methodology and the systems approach potentially provides that methodology. Since 1970 the systems approach has diffused with varying degrees of success into all areas of physical geography and as visualized by Stoddart (1967b, p. 538):

> Systems analysis at last provides geography with a unifying methodology and using it geography no longer stands apart from the mainstream of scientific progress.

Despite the widespread incorporation of the systems approach into branches of physical geography and its increasing use as a framework for textbook structures, some disillusionment has occurred, possibly because the systems approach did not immediately offer all that was at first hoped. It could not stand alone because it had to be applied to the traditional concerns of physical geographers as outlined in the four previous chapters. In turn however adjustments of attitude and method were demanded of the traditional concerns, some of which had developed simultaneously with, and sometimes interactively with, systems ideas. This chapter is therefore structured to show how systems thinking was incorporated into physical geography, to outline the way in which systems approaches have been utilized in other disciplines, and then to sketch the position of the several branches of physical geography in the early 1980s as a vehicle for proceeding towards what may be an emerging focus for a unified approach to the system as applied to physical environment at the scale traditionally employed by the physical geographer.

Developing a systems approach

The systems way of thinking was adopted successively by biogeography, soil geography, climatology and geomorphology and this adoption extended over about 35 years from 1935 to 1971 when *Physical Geography: A systems approach* (Chorley and Kennedy, 1971) was first published.

However the rate of incorporation of the ideas increased exponentially and was most dramatic in the decade from 1965 to 1975.

Ecosystem was a term proposed by the plant ecologist A.G. Tansley in 1935 as a general term for both the biome which was 'the whole complex of organisms – both animals and plants – naturally living together as a sociological unit' and for its habitat. Tansley (1946, p. 207) further expressed the notion that:

> All the parts of such an ecosystem – organic and inorganic, biome and habitat – may be regarded as interacting factors which, in a mature ecosystem, are in approximate equilibrium: it is through their interactions that the whole system is maintained.

In his review of organism and ecosystem as geographical models Stoddart (1967b, p. 523) showed how Tansley's concept broadened the scope of ecology beyond the purely biological content and gave formal expression to a variety of concepts covering habitat and biome which date back to the late nineteenth century. Fosberg (1963, p. 2) extended Tansley's definitions:

> An ecosystem is a functioning interacting system composed of one or more living organisms and their effective environment, both physical and biological. . . . The description of an ecosystem may include its spatial relations: inventories of its physical features, its habitats and ecological niches, its organisms, and its basic reserves of matter and energy; the nature of its income (or input) of matter and energy: and the behaviour or trend of its entropy level.

In relation to the island ecosystem Fosberg (1963, p. 1) avowed that:

> . . . . such partial concepts of nature as climate, vegetation, biota, soil environment, and even community, though very useful for analytical purposes, are not especially conducive to synthetic thinking or integration.

Although major developments in the ecosystem were largely external to biogeography until the 1960s, Stoddart (1967b, p. 524) argued that the ecosystem concept has four main properties which commend it for geographical investigation. First, it is monistic and brings together environment, man and the plant and animal world within a single framework in which interaction between the components could be analysed. Because emphasis is on the functioning and nature of the system as a whole this should dispense with geographic dualism. Secondly, the ecosystems are structured in a more or less orderly, rational and comprehensible way, and therefore provide an approach which requires identification of the structures present and the links between the structural components. Thirdly, ecosystems function as a result of through-put of matter and energy, and in ecology the identification of trophic stages and quantification of food webs and of productivity are exemplars of the way in which the function can be utilized. Fourthly, the ecosystem is a type of general system and therefore can be visualized as an open system tending towards a steady state under the laws of open system thermodynamics. Because ecosystems in a steady state possess the property of self-regulation this is analogous to mechanisms such as homeostasis in living organisms, feedback principles in cybernetics and servomechanisms in systems engineering.

The systems approach is now explicitly used in biogeography and Simmons (1978), for example, in discussion of the ecosystem scale distinguishes two approaches that are often made: one which is synoptic and develops from intuitive perception of an ecosystem to studies of ecological cohesion including, where relevant, the significance of human activity; and another approach which is more analytical whereby measurements are made of the flow and partitioning of energy through the ecosystem and of the cycles of mineral nutrients within the system. The latter approach embraces production ecology as study of the rate of production of organic matter in an ecosystem, and population ecology which studies changes in population numbers of the species of the ecosystem.

In soil geography it is generally suggested that the systems approach was formally applied by Nikiforoff (1959) although earlier he had distinguished accumulative and non-accumulative soils and so implicitly involved an open system attitude (Nikiforoff, 1949). Earlier still Jenny (1941) in his *Factors of Soil Formation* had expressed any soil character in terms of climate (cl), organisms (o), relief (r), parent material (p) time (t), and other additional unspecified factors in the relation:

$$s = f(cl\ o\ r\ p\ t\ \ldots)$$

In a later paper Jenny (1961) endeavoured to advance the approach by introducing ecosystems in a systems-oriented approach whereby the initial state Lo represents the assemblage of properties at time zero, Px is the combined result of inputs and outputs which provides the flux potential and t is the age of the system. This provided a revised general state factor equation where ecosystem properties (l), soil properties (s), vegetation properties (v) and animal properties (a) are combined in an equation in the form:

$$l, s, v, a = f(Lo, Px, t)$$

Thus five broad groups of factors were suggested:

$$l,s,v,a = \begin{cases} f(cl,o,r,p,t.\ldots) & \text{climofunction} \\ f(o,cl,r,p,t.\ldots) & \text{biofunction} \\ f(r,cl,o,p,t.\ldots) & \text{topofunction} \\ f(p,cl,o,r,t.\ldots) & \text{lithofunction} \\ f(t,cl,o,r,p.\ldots) & \text{chronofunction} \end{cases}$$

or a more convenient way of writing this could be in the form:

$$l,s,v,a = f(cl)_{o,r,p,t} \qquad \text{climofunction}$$

Although it is very difficult to analyse soil systems quantitatively using this approach, a number of attempts have been made to solve the state factor equations and these have been reviewed by Yaalon (1975). In the generalized theory of soil genesis proposed by Simonson (1959) there was a formal introduction of systems thinking into soil science with soil visualized as an open system with inputs and outputs.

These developments of the soil system were achieved within soil science but adoption and development of the approach in physical geography was achieved by Huggett (1975) when he extended the catena approach to the drainage basin, which he used as the basis for a model of the soil system, and attempted to simulate the flux of plasmic material in an idealized basin. This model is really an extension of the nine-unit landsurface model and adduces definable flow lines of material within soil-landscape units. Different constituents will move through the system at different rates and one time-step in the simulation could represent a day for mobile salts but as much a millenium for a relatively immobile element such as aluminium.

Climatology is similar to soil geography in that the incorporation of systems ideas were achieved by workers in disciplines other than physical geography but it is dissimilar in that there were few explicit statements of the systems approach. This was not necessary, however, in relation to the study of the atmosphere where the earlier emphasis upon climatic classification was gradually replaced by a shift towards the understanding of the surface energy and mass exchanges especially inland over vegetated surfaces (Hare, 1973).

Geomorphology received the clear import of general systems theory when Chorley (1962) clearly and explicitly reviewed the approach although acknowledging the earlier statements that had been made, particularly by Strahler who had avowed (Strahler, 1952, p. 935):

> Geomorphology will achieve its fullest development only when the forms and processes are related in terms of dynamic systems and the transformations of mass and energy are considered as functions of time.

and also by Hack (1960). This application (Chorley, 1962) emphasized the contrast between the open-system view which was commended and the closed system view which was at least partially embodied in Davis's view of landscape development. In a closed system the given amount of initial free energy becomes less readily available as the system develops towards a state with maximum entropy, where entropy signifies the degree to which energy has become unable to perform work. Open systems, however, were portrayed as those which need an energy supply for maintenance and preservation and as maintained in an equilibrium condition by the constant supply and removal of material and energy. Open systems can import free energy (or negative entropy) into the system and they can behave equifinally whereby different initial conditions can lead to similar end results. The value of the open-system approach to geomorphology was summarized (Chorley, 1962, p. B8) as depending upon the universal tendency towards adjustment of form and process; to direct investigation towards the essentially multivariate character of geomorphic phenomena; to admit a more liberal view of morphological changes with time to include the possibility of non-significant or non-progressive changes of certain aspects of landscape form through time; to foster a dynamic approach to geomorphology to complement the historical one; to focus upon the whole landscape assemblage rather than upon those parts assumed to have evolutionary significance; to encourage geomorphic investigations in those areas where evidence for erosional history may be deficient; and to direct attention to the hetereogeneity of spatial organization. A number of the implications were summarized as (Chorley, 1962):

> It is an impossibly restricted view, therefore, to imagine a universal approach to landform study being based only upon consideration of historical development. . . . the physical and the resulting psychological, inability of geographers to handle successfully the simultaneous operation of a number of causes contributing to a given effect has been one of the greatest impediments to the advancement of their discipline.

This very influential paper was succeeded by others, many also published as *Professional Papers*, and including exposition of the concept of entropy (Leopold and Langbein, 1962) which is referred to later (p. 159) although the emphasis upon contemporary environments which was being promulgated was reconciled to some extent with the more historical approaches by the time-scales devised by Schumm and Lichty (1965) as indicated at the beginning of chapter 8 (p. 162).

In the 1960s the import of the systems approach to geomorphology was probably reflected most significantly in the emphasis which developed in research investigations as many more dynamic investigations appeared. Not all geomorphologists were enamoured of the approach and Smalley and Vita-Finzi (1969) considered that general systems theory was unnecessary in the earth sciences and introducing confusion rather than clarification to empirical investigations. More generally in geography as a whole Chisholm (1967) dismissed the approach as formalizing what had been done before and employing 'a jargon-ridden statement of the obvious'.

Whereas the developments noted above had appeared in the context of branches of physical geography, in 1971 *Physical Geography A Systems Approach* was published (Chorley and Kennedy, 1971). Unlike previous texts, this book made an unreserved attempt to show how the phenomena of physical geography could be rationalized and perhaps given new significance and new coherence in terms of systems theory, and 'by avoiding the usual *pot pourri* of information about the earth and its atmosphere which had traditionally been termed physical geography' it was devoted to the identification and analysis of some of the more important systematic relationships with which modern physical geographers are concerned. It did not purport to deal with all the subject matter of the field and it was believed that:

> . . . ultimately the real value of this book may lie in the intellectual stimulation it

provides to view traditional material in physical geography in a new light.

The dual purpose of the book was identified as attempting to present a view of landscape and processes in terms relevant to the student of human geography by indicating the ways in which socioeconomic and physical systems interlock and interact, and also by showing how far knowledge of the physical world and its processes is compatible with the ideas of systems theory to demonstrate areas in which research might profitably be concentrated.

In achieving these aims it was undoubtedly enormously successful and must rank as one of the most cited and most influential physical geography textbooks of the twentieth century. Its impact could probably have been even greater but it ceased to become available to students by 1980 and this was unfortunate when many much less meritorious texts have soldiered on! The structure of the book was based upon the distinction of four types of medium-scale systems each showing distinct but complementary properties and giving a progressive sequence to higher levels of integration and sophistication. The four types of system, illustrated in Figure 7.1, were:

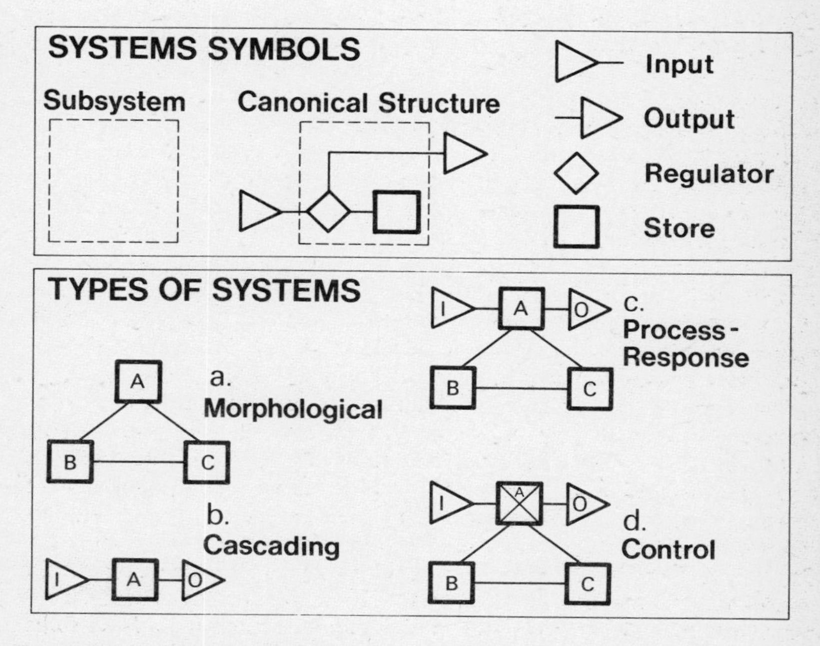

Figure 7.1: Systems terminology and classification (after Chorley and Kennedy, 1971). The four types of systems are summarized on p. 146.

a. *Morphological systems* which comprise morphological or formal instantaneous physical properties integrated to form a recognizable operational part of physical reality, with the strength and direction of connectivity revealed by correlation analysis.
b. *Cascading systems* composed of chains of subsystems which are dynamically linked by a cascade of mass or energy so that output from one subsystem becomes the input for the adjacent subsystem.
c. *Process–response systems* are formed by the intersection of morphological and cascading systems and involve emphasis upon processes and the resulting forms.
d. *Control systems* are those where intelligence can intervene to produce operational changes in the distribution of energy and mass.

The impact of this approach on physical geography can be attributed to the ways in which these four types of system could rationalize physical geography endeavour, and could catalyse the introduction of new concepts especially concerned with temporal change. In an evaluation of the book Cooke (1971) wrote of the work as:

> . . . a further contribution to geography's New Testament, which already includes several works by the senior author. Written with refreshing vitality and, perhaps, missionary zeal, it will delight the converted, and it may attract physical geographers who are earnestly seeking new solutions to certain of their difficulties.

Cooke noted how many of the examples were rather parochial in being drawn largely from drainage basin, slope and channel studies in geomorphology and hydrology and to a lesser extent from atmospheric research, so that the reader was left to determine how systems related to the biosphere and to the pedosphere. However in an assessment of the approach in 1971 Cooke (1971, p. 214) concluded:

> In terms of integrating physical geography with related disciplines, a systems approach undoubtedly succeeds. By adopting the terminology of systems, the physical geographer effectively translates his jargon into the *lingua franca* of much natural, physical and engineering science, and communication with these subjects is thus improved.

Subsequently applied to geography as a whole, Bennett and Chorley (1978) produced a text which attempted to explore firstly the extent to which systems theory provided an interdisciplinary focus for environmental matters and to what extent systems technology provides an adequate vehicle; and secondly to ascertain the manner in which systems approaches aid in the development of an integrated theory relating social and economic theory to physical and biological theory. The approach which they employed was by using hard systems which are capable of specification, analysis and manipulation in a more or less rigorous and quantitative manner; soft systems which are not tractable by mathematical methods; by exploring examples; and by combining interdisciplinary approaches by reviewing the dilemmas which confront man's intervention in natural systems. In their conclusion Bennett and Chorley (1978, p. 541) suggest that systems methods have illuminated thought, clarified objectives, and cut through the theoretical and technical undergrowth during the third quarter of the twentieth century in a most striking manner but in addition:

> At the very least we should view the systems approach as a complex training ground wherein to develop the ringcraft, the deftness of touch, the timing, the incisiveness of action, and the introspective forethought which are increasingly necessary in dealing with our escalating environmental problems. At most, the systems approach provides a powerful vehicle for the statement of environmental situations of ever-growing temporal and spatial magnitude, and for reducing the areas of uncertainty in our increasingly complex decision making control situations.

In an appraisal of the book by Bennett and Chorley (1978) and of a book applying systems throughout geography as a whole (Chapman, 1977) it has been suggested (Kennedy, 1979) that it is premature to adopt the systems approach exclusively if it would lead to the abandonment of other important parts of the subject.

Systems in science

When systems were being adopted in branches of physical geography there were similar developments in other earth sciences. The concept of a system was not new because Newton had written on the solar system, biologists had been concerned with living systems, and geography had implicitly used notions of the systems concept since early days of the subject. However as Harvey (1969, p. 449) noted, systems concepts, although old, had tended to remain on the fringe of scientific interest acting more as constraints than as subjects for intensive investigation. *General systems theory* was proposed by a biologist Ludwig van Bertalanffy (1901–) as an analytical framework and procedure for all sciences. He saw the theory as a way of uniting all sciences but the academic world did not readily accept the theory as presented at a philosophical seminar in Chicago in 1937 (Holt-Jensen, 1981). Although physics was concerned with general theory at that time the trend elsewhere was to emphasize detailed investigations and to avoid general theories. This attitude changed by the 1950s and was emphasized by the development of cybernetics, information theory and operations research. Cybernetics was a new branch of science in the late 1940s formed as a study of regulating and self-regulating mechanisms in nature and technology. It was primarily concerned with the control mechanisms in systems and with communication processes which determine their successful operation, and part of its mathematical basis is found in information theory (Holt-Jensen, 1981). The philosophical aspects of general systems theory that von Bertalanffy published in papers and books in the 1950s and 1960s could be approached using cybernetics. Bertalanffy (1972) distinguished three aspects of the study of systems. Firstly, systems science which deals with the scientific investigation of systems and with theory in various sciences. Secondly, systems technology which is concerned with applications in computer operations and theoretical development such as game theory. Thirdly systems philosophy which involves reorientation of thought and world view consequent upon the advent of system as a new scientific paradigm. A system has generally been defined as

(a) a set of elements with variable characteristics
(b) the relationships between the characteristics of the elements
(c) the relationships between the environment and the characteristics of the elements

In analysis, attention has been accorded to the structure of the system, its behaviour which involves energy transfer, its boundaries, its environment, its state whether transient or equilibrium, and its parameters which are unaffected by the operation of the system.

The systems approach has necessarily been identified with a positivist approach and as such has been less resilient in human (Johnston, 1983a) than in physical geography. Although it has been argued that most of the ideas central to general systems theory are certainly valuable, they had been applied to it without formal knowledge of the theory (Jennings, 1973, p. 124), nevertheless the theory has focused thinking and has probably been responsible for a more comprehensive view of many environmental situations. The theory is primarily inductive in nature and so lacks explanatory value but it may have helped to counter the trend towards specialization in science. Boulding (1968) has identified the crisis of science as arising from the fact that communication among disciplines and subdisciplines is increasingly difficult so that the greater the fragmentation into subgroups the more likely that the total growth of knowledge may be inhibited. Knowledge instead of being pursued in depth and integrated in breadth is pursued in depth but in relative isolation (Lazlo, 1972a) by the specialist who concentrates on detail and tends to ignore the broader context. The scientist taking a more generalist view should concentrate upon the structure and magnitude at all levels of magnitude, fit detail into the general framework and, by endeavouring to identify relationships, believe that some knowledge of connected complexity is preferable to an even more detailed specialized knowledge. One expression of this is by Medawar (quoted in Coffey, 1981, p. 30) that 'in all sciences we are progressively relieved of the burden of singular instances, the tyranny of the particular. We need no longer record the fall of every apple.'

A basic twofold hierarchical evolution was perceived (Lazlo, 1972b) in the many systems of the universe. Firstly, the entities of astronomy present a macrohierarchy embracing the entities of astronomy, galaxy clusters, galaxies, star clusters, stars, and planets and subsidiary bodies. Secondly in a microhierarchy are the terrestrial entities of physics, chemistry, biology, sociology and international organizations which are atoms, molecules, molecular compounds, crystals, cells, multi-cellular organisms and communities of organisms. The basic components of each hierarchy are atoms or their constituent elementary particles which are composed of quarks. Huggett (1976b), in an attempt to define geographical systems, extended Lazlo's (1972b) bipartite scheme to include another micro-hierarchy to provide an evolutionary link between atoms and planets in the macrohierarchy, namely the hierarchy of planetary and geological systems. Huggett (1976b) also noted that the systems of the atoms-to-planet and the atoms-to-societies hierarchies commingle to produce a third hierarchy which is the hierarchy of environmental systems. He proceeded to construe geography as the science that deals with systems at the uppermost levels of this environmental, or atoms-to-ecologies, hierarchy although noting that some geographers may delve into systems at a more detailed and fundamental level. Anuchin (1973) in his view of theory in geography has noted the uncertainty which has characterized much of the geography's growth and its attempts to become an established science when many scientists who were substantially geographers (he cites V.V. Dokuchaev as an example) have often preferred to give their activities another name and (Anuchin, 1973, p. 44):

> When one recalls that for decades geography was not recognized as a scholarly discipline, one cannot perhaps be surprised that some scientists have unsuspectingly spent their whole careers studying geography – rather like the well-known character of Molière who did not know that he had been speaking prose all his life!

Anuchin proceeded to style the subject matter of geography within the geographical sphere of the earth as a synthesis of all near-surface spheres (lithosphere, hydrosphere, atmosphere, biosphere, and the sociosphere or neosphere) into one interacting system.

In his application of systems analysis in geography as a whole Huggett (1980) has distinguished a strategy of systems analysis that is applicable to both theoretical, at either a micro- or macro-scale level, or an experimental mode of analysis which observes the nature of relationships between system parts. The strategy (Huggett, 1980) involved four phases identified as the lexical phase which necessitates identification of system components; the parsing phase which involves establishing relationships between system components; the modelling phase which requires expression of relationships in the context of a model and then calibrating the model; and the analysis phase in which there is an attempt to solve the system model and if not successful the procedure is repeated with a modified model.

An additional approach to geography as a whole using a systems framework has been advocated by Wilson (1981). Although a human geographer who has specialized in the analysis of urban systems particularly using entropy-maximizing models, in this book he recognizes three types of system and applies them to both physical and human geography. He characterizes systems analysis as being concerned with handling complexity, with identifying and understanding systemic effects, with seeking methods which are applicable to a wide range of systems classified into certain types, and with providing tools which aid planning and problem solving. He concludes that although developed during 25 years, progress has been very rapid, but in many fields there is still considerable scope for further empirical work.

Systems in physical geography

The conclusion of Wilson (1981) that a considerable amount of scope remains for the further empirical development of environmental systems is certainly exemplified by physical geography. The position achieved in the early 1980s can be assessed from the textbooks that have been written because these should be influencing the methods of instruction in the subject of physical geography as a whole, and also from the position in the branches of physical geography.

Physical geography books written for a variety of educational levels have adopted a systems viewpoint to provide an organizational structure. Thus in her *Physical Geography* King (1980a) adopts a systems framework in combination with an emphasis upon three scales of investigation and treats local-scale studies and continental-scale studies, and global-scale studies separately. In his book on *Environmental Systems* Dury (1981) adopts an approach to physical environment in which he adopts the four-fold classification of systems devised by Chorley and Kennedy (1971) but in addition adds a fifth type in the form of an ecosystem justified as including 'sets of biosystems which interact with one another and with their immediate physical environment'. In a sequel to his earlier well-established physical geography texts Strahler adopts a systems basis for his book in which (Strahler and Strahler, 1976):

> Two contemporary trends in physical geography are emphasized. One trend is toward a climatology geared more closely to the soil–water balance and the availability of soil water to plants. A second trend is toward linking physical and organic processes through the concept of cycling of energy and materials within ecosystems. To analyse problems of world food resources, the geographer must integrate his knowledge of physical systems with the ecologists knowledge of ecosystem dynamics.

In these examples, and more could be cited, the system is used as a convenient model for disseminating the contemporary attitude to physical geography. One book, published before Chorley and Kennedy, presented the geosystem as a single planetary system in which land, sea and air are dynamically integrated as a consequence of processes by which energy, matter and momentum are continually exchanged (Rumney, 1970).

Although this book was published before many of the systems texts, and therefore is largely non-quantitative, it was refreshing in that it was a brief treatment, which began with the energy budget of the atmosphere and proceeded to include all aspects of the environmental system including the sea in the geosystem, a topic frequently ignored by physical geographers in the second half of the twentieth century. Although ignoring man and control systems, taking many detailed facts from the period of physical geography before systems and beginning with two chapters which are very climatologically inspired, nevertheless the approach was refreshing in that it indicated what could be done to redress the imbalance detected by some physical geographers (e.g. Brown, 1975) and to counter the increasingly fissiparist tendencies of the previous decades.

Perhaps one of the greatest advantages of a systems viewpoint has been to cement the branches of physical geography more closely and therefore to make what Walton (1968) characterized as the unity of the physical environment a more realistic prospect. In branches of physical geography the systems approach was also used as part of a fundamental basis for the study of drainage basins, (Gregory and Walling, 1973) or of alluvial river channels (Richards, 1982) and it has been used as the basic vehicle for other branches of geomorphology. In presenting a geomorphological approach to glaciers and landscape Sugden and John (1976) utilize a simple systems approach as a vehicle for the explanation of complicated ideas because they believe 'in the value of a systems framework as a powerful explanatory tool'. In such cases the system is merely used as a collective vehicle for an open-system approach which involves input, storage, output relationships; for systems hierarchies; and for concepts such as thresholds and relaxation times which relate to change and are considered in chapter 8 (p. 174). More recently Sugden (1982) has utilized a systems framework for his synthesis of the character of the Arctic and Antarctic.

In geomorphology in general Thornes and Ferguson (1981) follow Weaver (1958) and Wilson (1981) in recognizing three kinds of systems. First are simple systems which involve not more than three or four variables and can be handled by relatively simple techniques, including regression models and partial differential equations although finite difference methods may increase in the future. Secondly are systems of complex disorder where there are large numbers of components and therefore of variables but only weak linkages between them, and they are handled by probabilistic methods of statistical mechanics. This includes probabilistic approaches to soil creep and to stream networks, coastal spit simulation and Box Jenkins models. Thirdly there are systems of complex order where there are a large number of components which have a strong organized interaction. The complexity increases as the square of the number of components so that simple analysis techniques cannot be employed and catastrophe theory and perturbation analysis are examples of appropriate analysis procedures.

In climatology the system has been adopted as providing a suitable framework and appears as the introductory foundation for *Causes of Climate* (Lockwood, 1979a), where it is argued that the application of systems theory and mathematics has completely changed the subject of climatology. The three types of systems which are distinguished are isolated systems which have boundaries which are closed to the import and export of both mass and energy; open systems where there is exchange of both matter and energy between the system and its environment, such as clouds; and closed systems in which there is no exchange of matter between the system and its environment though there is in general an exchange of energy. The atmosphere, oceans and landsurfaces are considered as a series of cascading systems connected by flows of mass or energy. In a review of climatology as it relates to geography Terjung (1976) reviewed the position of climatology in geography teaching and research and suggested:

> . . . geographers who are interested in the physical environment of man need to be trained rather differently than in the past. We can afford no second-class investigators. Geographers who want to work in the environmental sciences must be willing to learn the methods of these sciences, lest we lose the respect of our colleagues in such fields. Prospective physical geographers should take basic courses in calculus, physics, chemistry, engineering, modern biology, and computer programming. On higher levels of instruction, geography departments should develop courses which stress a core of basic thermodynamics and hydrodynamics and their relation to the environmental envelope of relevance to mankind. Instead of drumming the trivial facts of physical geography of yesteryear into the heads of reluctant freshmen, introducing classes should teach the concepts of systems analysis and the flows of energy, mass, momentum, and information through various environments of our planet.

Acknowledging the creative ideas in the systems approach of Chorley and Kennedy (1971) five classes of methodology were identified by Terjung (1976). These include qualitative inventory and associations; quantitative structural correlations analogous to the morphological system in geomorphology; functioning processes concentrating upon the paths followed by energy, mass and momentum through the subsystems of the planetary boundary layer, the interfaces near the earth's surface, and the soil–plant–water systems; physical process–response systems which link the morphological and cascading systems; and finally physical–human process–response systems which are equivalent to the geographical control systems identified by Chorley and Kennedy (1971). Although this fifth category is as yet insufficiently developed Terjung (1976) argued that it is increasingly appropriate as a research level for geographer climatologists.

In hydrology the systems approach has been easily assimilated because the hydrological cycle readily lent itself to systems representation and because use of the cycle, as in water resource systems, further commended the approach. Therefore a number of general geographic texts (e.g. Bennett and Chorley, 1978; Huggett, 1980; Wilson, 1981) have used illustrations from this subject area and such illustrations were also prominent in physical geography (Chorley and Kennedy, 1971). Approaches to hydrological model building have been either via physical hydrology, which is the investigation of the components of the hydrological cycle to achieve a full understanding of the mechanisms and interactions involved, or by systems synthesis investigations, which attempted a complete simulation of drainage basin operation by adjusting the components and the parameters of the model until outputs from the model agreed with empirical results from known inputs. Physical geographers have been most devoted to the physical hydrology approach using models for subsystems such as evaporation, infiltration, surface runoff or groundwater or overall catchment models. Although physical geographers have contributed less extensively to the optimization of general systems synthesis models, nevertheless they have to be aware of the work achieved by engineers. More (1967) in a review of hydrological models in geography concluded that although there are large areas of overlap between hydrology and geography, the two disciplines have developed quite separately, that many of the hydrological models developed have not been geographic in their inception, but that geographers should not ignore the implications of the fast-moving science of hydrology. When investigating the hydrological system a convenient distinction has been made (Amorocho and Hart, 1964) between parametric hydrology which develops relationships between physical parameters involved in hydrologic events, and stochastic hydrology which employs the statistical characteristics of hydrologic variables to solve hydrologic problems.

Soil and vegetation systems have been considered together by Trudgill (1977) when he perceptively noted that:

> The maddening thing is that once it is realized that the environment is a mutually reactive system, it becomes extremely difficult to isolate and discuss specialized points in a simple way without at the same time being tempted to discuss every interaction and interrelationship which exists in the system.

The expedient adopted by Trudgill is to deal with each component of soil and vegetation in turn and then to build up a sequential picture of the whole system. The components of nutrient systems treated are weathering and atmospheric inputs, the leaching output and nutrient cycling, and this leads to models of the system and to models of stability and change. Although Trudgill (1977) did not follow his initial intention to deal in detail with the three main systems of flow and cycling nutrients, energy and water, these energy flows are prominent in his book. In his alternative biogeography Gersmehl (1976) also focuses upon the circulation of mineral elements modelled as a system of compartments and transfer pathways.

A focus within the environmental system

A review of the way in which a systems viewpoint has been adopted by physical geography textbooks as well as in some research is justified because the alacrity with which the approach was adopted perhaps emphasizes that there was a clear need for a unifying approach. The systems viewpoint has certainly retarded, or perhaps even reversed, the trend towards great specialization and separation of the branches of physical geography from each other and from human geography. However the adoption of the systems approach as a textbook framework, although producing some stimulating texts which effectively transcended the traditional boundaries of physical geography (e.g. Chorley and Kennedy, 1971; Trudgill 1977), has not been paralleled by an equal amount of adoption as a vehicle for research programmes. Is there a danger that physical geographers pay lip service to the benefits of the systems approach but do not adopt, adapt and apply it sufficiently in their researches? Perhaps also in certain textbooks the systems approach has become adopted in outline as a prelude to conventional content without requiring any fundamental adjustment in the content of the physical geography textbooks. It is important to distinguish between systems analysis which can mathematically optimize some attribute of the links within and between systems, and General Systems Theory which purports that all systems can be understood by the application of systems principles. The danger of adopting a systems approach uncritically is that it is assumed that it is sufficient to identify system structures and to portray the multitudinous variables involved in a particular system which then reinforces the first law of ecology as graphically described by Commoner (1972) that everything is connected to everything else. However Commoner (1972) also stated three other laws of ecology; everything must go somewhere; nature knows best; and there's no such thing as a free lunch because somebody somewhere must foot the bill. Perhaps physical geography has been overtly concerned with the parallel to the first law rather than proceeding to the equivalent of the other three!

A further example of the somewhat static way in which physical geographers have tended to utilize a system is afforded by the runoff-producing system. Many physical geography texts and papers have used the hydrological cycle as an exemplar of the way in which systems components may be identified and the structure expressed. However although the diagrams presented identify the stores and the pathways between the stores they do not effectively represent the dynamics of the runoff-producing system. Such dynamics have to be visualized in three dimensions because, as the area of saturation adjacent to water courses expands during a storm, so the runoff-producing area expands and hence other types of runoff begin to influence the operation of the system. Thus a systems diagram cannot completely represent the dynamic situation and the way in which changes occur during a storm event.

To progress beyond such a static application of systems it is necessary to focus upon the dynamics of the system and in the preceding few pages a number of instances have been cited where physical geographers have already advocated placing emphasis upon the function of the system. These include the approaches adopted by Lockwood (1979a), by Terjung (1976) and by Simmons (1978), the approach considered by Trudgill (1977) and the approach developed in an introductory text by White, Mottershead and Harrison (1984). In each of these cases the emphasis is placed upon the transfers of mass and energy and this may provide a much-needed focus of further development for the extension of the systems approach employing the notions of rate of doing work or of power. Such transfers of mass and energy are fundamental features of cascading systems as recognized by Chorley and Kennedy (1971) but are capable of further emphasis.

The geography of energy was the subject for a paper by Linton (1965) which appeared well before the advent of the systems approach. Although Chorley (1973b, p. 157) argued that Linton's idea of:

> . . . flows of capital investment, population, technological information, generated energy, water, and the like, together with such constraints as involve interest policies and the mechanisms of group decision-making, can be reduced to comparable units so as to be structured into energy linkages similar to those of ecosystems is clearly an illusion. . . .

however in physical geography the geography of energy may afford a pertinent focus. Considering geography as the description of the changes that take place or have taken place in or at the surface of the earth Linton (1965) suggested that any changes which occur in the real world imply that work has been done and energy expended. Four sources of energy were identified namely radiant energy from the sun; internal energy from the earth's interior; rotational energy of the whole and parts of the solar system; and vital energy which is energy in the service of man. Each of these sources was explained very geographically and Linton (1965, p. 227) concluded:

> To my academic colleagues on both the physical and human sides and the biogeographical middle of the subject I would like to express the hope that my method of expressing salient parameters, in fields as far apart as climatology and social geography, in terms of a common set of units – the Watt and the calorie – has value for the future of our subject.

Perhaps the future which Linton perceptively visualized has not yet been realized in this way but the basis of energy has certainly been advocated in the branches of physical geography, and perhaps most clearly in climatology. In the paper which succeeded that by Linton, Hare (1965) emphasized energy exchanges within the atmosphere where energy signified the capacity to do work and where work is what happens when force accelerates mass over distance. Subsequently Hare (1966) suggested that the outstanding change in climatology in the post-1945 years has been the shift away from parameters such as temperature and relative humidity and towards the measurement of fluxes. This has necessitated a concern with the movement and transformation of energy in the atmospheric boundary layer, in the plant cover and in the soil so that progress could be made towards understanding the mechanism of energy and moisture exchange. Subsequently it was suggested (Hare, 1973) that this trend in climatology had been emphasized by three other contributions identified as the micrometeorological method whereby a branch of experimental physics has used experimental techniques to provide physical insights into the nature of the earth's surface and planetary boundary layers; the microclimatological method where the techniques of micrometeorological measurement and boundary layer theory together with the related parts of soil physics and plant physiology are applied to exchange and transformation processes over natural surfaces, leading to study of energy transformation at the land surface, water movement upwards and downwards through the soil, and to how carbon dioxide is assimilated during photosynthesis and released during respiration; and the hydrological method whereby energy exchange during the hydrological cycle was quantified. These methods are analogous to the energy-centred methods adopted by ecologists. In his view of climatology for geographers Terjung (1976) has also advocated the emphasis on the process–response systems related to mankind which occur with the planetary boundary layer, interface and substrates and he has advocated study of flows of energy, mass, momentum and information through the various environments of the planet earth (see p. 152). Similarly in their view of climatology: 'The challenge for the eighties' Mather *et al.* (1980) concluded that climatology must systematically investigate the exchanges of heat, water and momentum that occur at or near the earth's surface and should focus upon topoclimatology as well as on transfer processes.

In a search for a more unifying theme for the physical–human geography interface, Simmons (1978) advocated the study of energy in contemporary society. Although not referring to Linton's earlier (1965) article published in the same journal, Simmons did trace the relevance of the Lindeman model of ecosystems and its development in a book by H.T. and E.C. Odum (1976) on *Energy Basis for Man and Nature.* This book begins with the statement that everything is based on energy, provides simple box models of energy flows in ecosystems and so embraces both cultural and natural components. In addition to clear definitions of energy, power as the rate at which energy flows, and efficiency which is any ratio of energy flows, Odum and Odum (1976) propose three principles of energy flows; the law of conservation of energy; the law of degradation of energy which introduces entropy as a measure of technical disorder to signify the extent to which energy is unable to do work; and the principle that systems which use energy best survive, which is the maximum power principle or minimum energy expenditure principle. Adopting the taxonomy of energy systems employed by ecologists Simmons (1978) suggests that geographers could use a set of ecosystem types which broadly provide a set of spatial regions (Table 7.1) which conform to patterns identified from analysis of

Table 7.1: Ecosystem types according to energy level (Based upon Simmons, 1979 after Odum, 1975)

	Ecosystem type	Annual energy flow *kilocalories per sq metre (kcal/m^2/yr^{-1})* Range	Average (estimated)
1	Unsubsidized natural solar-powered ecosystems, e.g. open oceans, upland forests. Man's role: hunter-gatherer, shifting cultivation	1 000-10 000	2 000
2	Naturally subsidized solar-powered ecosystems, e.g. tidal estuary, lowland forests, coral reef. Natural processes aid solar energy input: e.g. tides, waves bring in organic matter or do recycling of nutrients so most energy from sun goes into production of organic matter. Man's role: fisherman, hunter-gatherer.	10 000-50 000*	20 000
3	Man-subsidized solar-powered ecosystems. Food and fibre producing ecosystems subsidized by human energy as in simple farming systems or by fossil fuel energy as in advanced mechanized farming systems. Green Revolution crops are bred to use not only solar energy but fossil energy as fertilizers, pesticides and often pumped water. Applies to some forms of acquaculture also.	10 000-50 000*	20 000
4	Fuel-powered urban industrial systems. Fuel has replaced the sun as the most important source of immediate energy. These are the wealth-generating systems of the economy and also the generators of environmental contamination: in cities, suburbs and industrial areas. They are parasitic upon types 1–3 for life support (e.g. oxygen supply) and for food; possibly fuel also although this more likely comes from under the ground except in less developed countries where wood is still an important domestic fuel.	10 000-30 000 000	2 000 000

*The most productive natural ecosystems and the most productive agriculture seem to have upper limits of *c* 5 000 kcal/m /yr.

satellite data. Subsequently in casting his text on biogeography Simmons (1979a) imaginatively used energy as an early key to the understanding of natural biogeography which through food chains, productivity, nutrient cycles and population dynamics provides the basis for the subsequent treatment of cultural biogeography. Elsewhere it has been argued that ecosystems are ordered arrangements of matter in which energy inputs carry out work (Stoddart, 1965) and that if the energy input is removed the structure will break down until the components are randomly arranged which with maximum entropy is the most probable state. Stoddart subsequently noted (1967b, p. 537):

> Geography is clearly concerned with systems on a multitude of levels. A preliminary attempt to develop a science of 'geocybernetics' has been made in a little-known paper by Polonskiy (1963). . . . The study of geosystems may now replace that of ecosystems in geography. . . .

A focus upon energy flows is certainly appropriate for physical geography and Simmons (1978) has considered how it may provide a more generally applicable theme but has cautioned (Simmons, 1978, p. 320):

> A case can thus be made for energy flows as linkages between man and environment, both in terms of resource uses and environmental impact. But caution is necessary, for the homogeneity of kilocalories and gigajoules may hide qualitative and cultural aspects of the flows which as geographers we are not at liberty to ignore. F.E. Egler sounded the same note when talking about the way ecological energetics – especially the trophic level concept – ignored taxonomic consideration. He said that ecological energetics was like grinding up cows to make hamburgers – you could not be sure a monkey had not slipped in somewhere.

The basis for use of energy in biogeography derives not only from work in ecology generally but also from the field of bioenergetics. Thus Broda (1975) has suggested that until the significance of the first law of thermodynamics that heat is a form of energy, had been appreciated, and until energy distribution implications of the second law had been assimilated, then progress in the analysis of bioenergetical processes was uneven and patchy. This can be developed by regarding organisms as chemodynamical machines, by identifying three classes of bioenergetic processes namely fermentation, photosynthesis, and respiration, and by utilizing an approach to classification based upon microphysiology and biochemistry as well as upon macrostructure of organisms and upon macrophysiological processes. Broda, who was in 1975 in the Institute of Physical Chemistry in Vienna, then applies bioenergetics to early conditions on earth and to the ecosphere in general. Although it could be assumed by some that this is a development towards a more specialized and fragmented approach it may alternatively be a foundation for a more unified approach and one which accommodates the necessity to combine understanding at the microphysiology and biochemistry (realist?) level with that at the biome level which had traditionally been approached more in functional terms.

In soils energy considerations have already been noted (p. 142) and Gerrard (1981) referring to Runge (1973) has reviewed the energy status of soil systems by recognizing three components. The decay component is where the energy status gradually declines and eventually the system should continue to a state of virtual exhaustion; the cyclic component occurs because energy and possibly material input changes in a rhythmic manner associated with diurnal and seasonal climatic cycles; and a random component is provided by irregular supplies of energy such as rainstorms. In the Runge (1973) model soil development is visualized in terms of organic matter production, time, and the amount of water available for leaching. The water available depends upon the amount of water that infiltrates and becomes available for pedogenesis compared with that which is removed by surface flow and hence is not available. The three-dimensional view of the plasmic flux of material in an idealized basin is modelled by Huggett (1975).

In hydrology, definition of the pathways of movement through the hydrological cycle has necessarily provided an important focus in studies by physical geographers but in geomorphology the use of concepts based upon energy-balance concepts has been less striking. Hare (1973) has suggested (see also p. 111) that this is because fluvial processes tend to be dominated by extreme events rather than balance relationships, and that the geomorphic time-scale is very long compared with that appropriate for climatic processes, although glaciology is in a very different situation. Indeed it is in the field of glacial geomorphology that one of the most imaginative approaches has been devised by Andrews (1972), who provided an analysis of total glacier power (W_T) as the product of basal shear stress and the average velocity. Effective power (W_E) was determined by the proportion of the total average velocity resulting from basal sliding so that the ratio W_T/WE could vary largely according to the proportion of basal slip to internal ice deformation. Andrews proposed that W_T/WE is small, between zero and 0.2 for polar and sub-polar glaciers, but between 0.5 and 0.8 and tending towards 1.0 for temperate glaciers. The implication which follows, that the glacial erosional forms produced by arctic and by temperate glaciers differ in size and geometry, receives some support from the glacial geomorphology literature. A further application in geomorphology was developed by Caine (1976) when he estimated the physical work in joules represented by different types of sediment movement. More generally, energy together with forces, resistances and responses, is used to introduce process in geomorphology by Embleton and Thornes (1979) where energy is attributed to solar radiation atomic energy, chemical energy, gravity, and energy of the earth's rotation sources.

Energy therefore provides a potentially useful theme throughout the branches of physical geography and there are indications that it could be the basis for a more integrated approach. Perhaps the most important and fundamental paper was by Leopold and Langbein (1962) when they reviewed the concept of entropy in landscape evolution. Because entropy relates to the distribution of energy the principle was introduced that the most probable condition exists when energy in a river system is distributed as uniformly as possible according to the physical constraints. This principle which is analogous to the implication of the second law of thermodynamics in relation to thermal energy, governs the energy transfers in fluvial processes, the spatial relations at any one time, and the sequence of development from one stage in geomorphic history to another. By developing the principle of least work as one of several ways in which the condition of maximum probability may exist in relation to entropy, Leopold and Langbein (1962) theoretically derived the longitudinal profile of rivers and the hydraulic geometry of river channels. A more general approach to the physical-geographical sphere and the anthroposphere has also been offered using entropy in relation to a cybernetic system by Krcho (1978). This structural attempt, although not referring to earlier work in English by Leopold and Langbein (1962) or Chorley and Kennedy (1971) affords one indication of the way in which a general energy-based approach is being sought.

Although already referred to elsewhere (p. 116) the most generally applicable energy-based approach is perhaps that advocated by Hewitt and Hare (1973). Proceeding from the system of exchanges of energy and mass in the biosphere and between the atmosphere and land surface, they show how the functions of an ecosystem require a never-ending series of exchanges of energy, water, atmospheric gases and mineral nutrients between the organic and inorganic parts of the system. Although models of paths and storage reservoirs have been developed by ecologists, geochemists and climatologists, there is a need to progress towards a multidimensional model of the entire system and in relation to physical geography Hewitt and Hare (1973, p. 37) conclude:

> The vast volume of writing on environment in the past five years has produced a cloud of obfuscation, but here and there among it one can learn from one's scientific and political neighbours.

Perhaps in the import of systems, more than in any other development in twentieth-century physical geography, there has been a need to learn from such scientific and political neighbours. Perhaps too the future of systems, by focusing upon the mechanics of the system, will necessarily incorporate processes, man and change over time, and employ quantitative methods – thus providing an integration of the themes in the previous four chapters. This has been achieved in outline in the urban system by Douglas (1983). However, systems change and the more integrated approach and understanding has led to the inception of new and original concepts and these will provide the basis for chapter 8.

Part III Trends in ten years 1970–1980

8

Time for change

Whereas the application of the systems approach in physical geography has been effected in several countries and received considerable stimulation from work in other disciplines, impressive new approaches to the study of time particularly in geomorphology and especially in fluvial geomorphology arose in North America. In the preceding chapters a number of characteristics of studies of process (p. 87), of investigations of chronology (p. 65) of assessment of human activity (p. 116) and of the applications of systems approaches (p. 140) have all proceeded to lead towards an improved understanding of, and basis for investigation of, temporal change. Contrary to the fears of some physical geographers therefore, these are the basis for a more integrated approach to the branches of physical geography and also for closer liaison between the branches themselves. Like earlier chapters, this chapter does not do effective and equivalent justice to all of the branches of physical geography: it concentrates upon water in relation to the study of temporal change, but this may be viewed as an example of new attitudes, concepts and approaches that are occurring in other branches of physical geography as well.

Perhaps most physical geographers would agree, however, that one paper stands out not just in geomorphology but in physical geography as a whole and that is the seminal paper by Schumm and Lichty in 1965 entitled Time, Space and Causality in Geomorphology. This paper (Schumm and Lichty, 1965) is certainly one of the most, if not the most, frequently cited geomorphological paper during the 20 years 1965 to 1985. It is a paper which initiated a new attitude to temporal change by proposing a method of reconciling views which had hitherto conflicted. In particular the dynamic equilibrium view advanced by Hack (see p. 97) and studies which emphasized investigations of process appeared to be at variance with investigations of long-term environmental change. More precisely, the approach of dynamic equilibrium as applied to landforms could cast doubt on the existence of accordant summit levels as reflections of formerly more extensive planation surfaces. Because Schumm and Lichty saw the possibility of misunderstanding time in geomorphic systems they stated that (Schumm and Lichty, 1965, p. 262):

> We believe that distinctions between cause and effect in the moulding of landforms depend on the span of time involved and on the size of the geomorphic system under consideration. Indeed as the dimensions of time and space change, cause–effect relationships may be obscured or even reversed, and the system itself may be described differently.

At least two original ingredients account for the substantial impact which this paper made. First, the distinction of three separate time-scales namely cyclic or geologic time, which encompasses millions of years as required to complete an erosion cycle; graded-time, which may be hundreds or thousands of years during which a graded condition or dynamic equilibrium exists; and steady-time typically of the order of a year or less when a true steady-state situation may exist. Second was the accompanying explanation of the status of geomorphic variables according to the time-scale being investigated. Thus a variable which is dependent at one time-scale may be independent at another and this aspect of status was illustrated by a well cited table (Table 8.1) which illustrates the notion with reference to drainage basin variables. Schumm and Lichty (1965, p. 266) drew attention to the fact that landscapes can be considered either as a whole or in terms of their components, and they can also be considered either as a result of past events or as a consequence of contemporary processes so that:

> Depending on one's viewpoint the landform is one stage in a cycle of erosion or a feature in dynamic equilibrium with the forces operative. These views are not mutually exclusive. It is just that the more specific we become the shorter is the time span with which we deal and the smaller is the space we can consider (Schumm and Lichty, 1965, p. 266).

The authors thus reconciled viewpoints that had hitherto appeared as alternatives and this is why the viewpoints were each accorded a separate chapter (4 and 5) in Part 2 of this book. The very stimulating paper by Schumm and Lichty was the first of a number of approaches to time which are outlined in this chapter. Some of the more specific achievements in the branches of physical geography are included and the ways in which interaction between the branches of physical geography is developing are introduced and one of the directions in which this leads is towards greater interdisciplinary awareness, discussed in the final section.

Views of time

In *Fluvial Processes in Geomorphology*, a book which had a very significant influence upon the movement towards the investigation of landscape processes, it was argued (Leopold, Wolman and Miller, 1964, pp. 7–8):

Table 8.1: A The status of river variables during time spans of decreasing duration (*after Schumm and Lichty, 1965*).

River variables	Status of variables during designated time spans		
	Geologic	Modern	Present
1 Time	Independent	Not relevant	Not relevant
2 Geology	Independent	Independent	Independent
3 Climate	Independent	Independent	Independent
4 Vegetation (type and density)	Dependent	Independent	Independent
5 Relief	Dependent	Independent	Independent
6 Palaeohydrology (long term discharge of water and sediment)	Dependent	Independent	Independent
7 Valley dimensions (width, depth, slope)	Dependent	Independent	Independent
8 Mean discharge of water and sediment	Indeterminate	Independent	Independent
9 Channel morphology (width, depth, slope, shape, pattern)	Indeterminate	Dependent	Independent
10 Observed discharge of water and sediment	Indeterminate	Indeterminate	Dependent
11 Observed flow characteristics (depth, velocity, turbulence, etc.)	Indeterminate	Indeterminate	Dependent

B The status of drainage basin variables during time spans of decreasing duration

Drainage basin variables	Status of variables during designated time spans		
	Cyclic	Graded	Steady
1 Time	Independent	Not relevant	Not relevant
2 Initial relief	Independent	Not relevant	Not relevant
3 Geology	Independent	Independent	Independent
4 Climate	Independent	Independent	Independent
5 Vegetation (type and density)	Dependent	Independent	Independent
6 Relief or volume or system above base level	Dependent	Independent	Independent
7 Hydrology (runoff and sediment yield per unit area within the system)	Dependent	Independent	Independent
8 Drainage network morphology	Dependent	Dependent	Independent
9 Hillslope morphology	Dependent	Dependent	Independent
10 Hydrology (discharge of water and sediment from system)	Dependent	Dependent	Dependent

Detailed understanding of geomorphic processes is not a substitute for the application of basic geologic and stratigraphic principles. Rather, such understanding should help to narrow the range of possible hypotheses applicable to the explanation of different geomorphic forms and surficial earth processes and deposits. . . . field investigations of modern process cannot be segregated completely from historical aspects of landform development.

Evolution of environments over time has of course been a long-established component of environmental research echoing the geological tradition which was basic to much early physical geography. More recently Chorley and Kennedy (1971, p. 251) highlighted the distinction between timeless and timebound changes although the distinction is not sustained by all scholars who regard being and becoming as aspects of an identical process which are limited by behaving. To illustrate this notion Chorley and Kennedy (1971) used a spiral (Fig. 8.1) in which cause and effect pass from

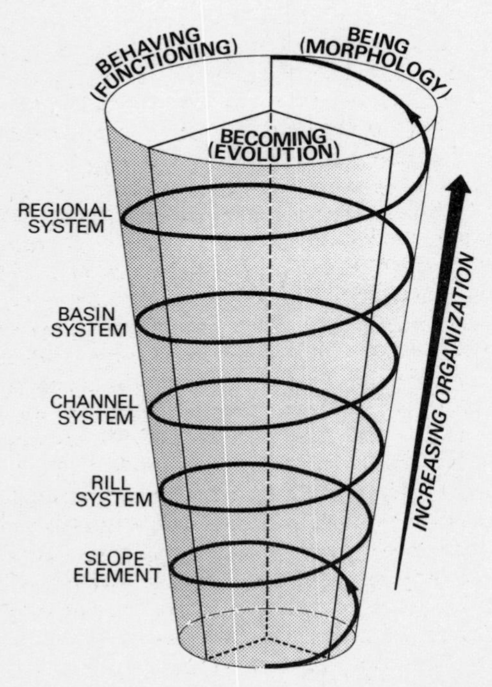

Figure 8.1: Systems architecture in time. Developed from a diagram used by Chorley and Kennedy (1971, Figure 7.1) based upon Gerard (1964).

becoming at one level of integrative organization to *being* when adopting a characteristic structure or morphology at a higher level, to *behaving* by adopting a characteristic morphology or structure at a still higher level. In this view the progressive integration of organization within the system leads inevitably to a sequence of evolution which is irreversible. In investigating time it has often been necessary to employ the ergodic hypotheses whereby space is substituted for time and an early illustration quoted by Chorley and Kennedy is that provided by Savigear's (1952) study of the profiles of sea cliffs in South Wales where cliffs have progressively been protected from basal marine erosion as the Laugharne Burrows spit has been extended eastwards. The 17 cliff profiles could then be arranged in a time sequence providing an excellent example of the substitution of sampling in space for sampling in time. The study of temporal change has also necessitated analyses of temporal sequences of data where these are available and a variety of historical methods has been utilized (e.g. Hooke and Kain, 1982) and where data sequences are not readily available it has been possible to utilize simulation to demonstrate the ways in which change may occur under closely controlled conditions. Wave tanks have been used for the simulation of coastal processes, and it has also been possible subsequently to simulate change in mathematical or computer models and a computer-based simulation model of Hurst Castle Spit on the southern coast of England was described by King and McCullagh (1971). Because the analysis of observed or simulated trends is fundamental to the investigation of temporal change, methods have been sought for the analysis of temporal data. Three types of serial dependence were identified by Thornes and Brunsden (1977) as including trend, periodicity and persistence. Trend is the long-term pattern with the short-term variations excluded by a filtering technique, periodicity occurs because of regular periodic or cyclic fluctuations which could be controlled by climate for example, and persistence arises either due to the presence of physical factors which produce the output series or as a consequence of the collection and manipulation of the data set. When data are compared from two or more series it has been necessary to adopt time-series statistics.

The objectives of the study of temporal change in the physical environment remain as dependent essentially upon understanding the past and estimating the future. Until 1970 temporal change was investigated with reference to long-established and often qualitative models which had insufficient basis in quantitative measurements or contemporary environmental processes. With the acquisition of more data and understanding of contemporary processes it has been possible to progress towards the development of more sophisticated models of temporal change and these have been of great significance in contributing to recent progress in physical geography. This can be illustrated by two or more series which cover the same period when time-series statistics have been used to establish the autocorrelation between the two series. This has allowed advances to be made beyond the process–response systems often analysed by correlation and regression analysis and in relation to application in physical hydrology Anderson (1975) has concluded that:

> Time series and related models enable an insight to be gained into the structure of comparatively long-term responses and occasionally of the whole basin system, and as such merely provide the source of one possible standard against which comprehensive process–response models of the basin hydrologic system can be judged.

In cases where series are themselves arranged in series and can be regarded as inputs and outputs to systems it has been possible to develop the mathematical function which relates the inputs and outputs and such a relation is known as a transfer function. In hydrology the drainage basin system is the basis for establishing a relation or transfer function between precipitation input and discharge output. Thornes and Brunsden (1977) illustrate this approach and argue that although not widely applied in geomorphology, transfer function modelling seems destined to become more widely utilized because geomorphologists and hydrologists are concerned with storage. Other developments pertinent to temporal change can be considered after reviewing developments in some of the branches of physical geography.

Time and the branches of physical geography

As climate is a fundamental control upon environmental systems the study of climatic trends is vital for the investigation of change. The climatic record of the post-glacial is of significance to several branches of physical geography and it has been demonstrated how temperature ameliorated in middle and high latitudes after the decay of continental ice sheets. This amelioration included a thermal maximum between 5000 and 3000 BC when summer temperatures were several degrees higher than at present, followed by a decline when cold wet conditions obtained in Europe around 900–500 BC (Barry and Chorley, 1976). Subsequently there was a warmer period in many parts of the world between about AD 1000 and 1250, and a deterioration between AD 1550 and 1700 is usually known as the Little Ice Age. More recently there has been a warming trend evident in glacier margin retreat, in a rise of the snowline, in a rise of sea level, and in northward recession of the tundra margin. Although most recently there has been indication of a halt to this trend or possible cooling, this is complicated by the influence of human activity which through the influence of increased CO_2 may be leading to increased temperatures. Long-term climatic changes have been reviewed by Lockwood (1977) with particular relation to the work in physical climatology that has focused upon causes and effects of variations in the solar constant. A simple mathematical model of the general atmospheric circulation was employed by Wetherald and Matanabe (1975) to explore the effects of changes in the solar constant and this indicated that the most dramatic effects were on the hydrological cycle, where a 2 per cent increase in the solar constant could cause up to a 10 per cent increase in the rate of precipitation and therefore of evaporation. This theoretical treatment was compared by Lockwood (1977) with simulations of ice-age conditions achieved by CLIMAP project members (1976) and by Gates (1976) which mapped boundary conditions such as sea surface temperature and surface albedo and demonstrated the differences between July ice age conditions 18,000 BP and those of the present. In the northern hemisphere surface air temperature could have been 5.3°C lower, and precipitation – 1.2 mm day^{-1} less (Gates, 1976).

Such promising research developments, although not necessarily the results of research by physical geographers, nevertheless have considerable significance for physical geography. The themes that have been highlighted in recent climatic research include, *firstly* the causes of climatic fluctuations (Lockwood, 1979b) and the advances in climate theory (Barry, 1979) where energy budget models are promising but require adequately distributed world data to refine their performance. *Secondly* the increased use of models may enhance the reconstructions of climates for geological periods. Particularly for the Quaternary it is established that there is a connection between the earth's orbit and ice ages but the exact nature of the connection between the long-term periodic variations in the earth's orbital parameters, as embraced by the Milankovitch theory, and the occurrence of glacial–interglacial transitions requires elucidation (Lockwood, 1980). *Thirdly* it is short-term fluctuations in climate that have attracted considerable interest by physical geographers, and this extends from the last 75–100 years, when the pattern of global trends is still difficult to isolate because of major gaps in the world data network especially over the oceans (Barry, 1979), to consideration of shorter periods because on a global scale variability of weather is more pertinent to food production than are climatic trends. Therefore over areas like the United Kingdom the frequency of occurrence of inland flooding could be affected by changes in the frequency of occurrence of intense precipitation (Perry, 1982). From analysis of the frequency of major flood events in the middle and lower Swansea valley 1875–1981 it was concluded that flood magnitude-frequency in the Tawe and Ebbw valleys has increased since the late 1920s and that the main reason is a marked increase in the magnitude-frequency of heavy rainfalls in South Wales and especially in the southern part of the Brecon Beacons uplands since the late 1920s (Walsh, Hudson and Howells, 1982). Particularly significant in consideration of recent climatic fluctuations has been the influence of vegetation on climate where it has been shown that changing the vegetation type can alter the regional climate (Lockwood, 1983b); the effects of the southern oscillation, which is a variation in pressure between the eastern and western Pacific over a period of years and the anomalously warm sea surface temperature called El Niño off the coast of Peru (Lockwood, 1984); and the importance of CO_2. Writing in 1984 Bach (1984) contended that although the acid rain problem was then in the forefront of world environmental and political attention it may transpire in the long run that the CO_2 climatic risk may have more far-reaching consequences for society. Thus the attention directed to short-term climatic fluctuations is projected as a basis, *fourthly*, for estimation of the climatic future. Weather forecasting is a long-established objective for the meteorologist but climate forecasting has now come to be regarded as of high priority and the first objective of the Global Atmospheric Research

Program (GARP) was to achieve greater understanding of short-term weather processes and thence of improved forecasting.

Climatic change and climatic fluctuations have thus become the subject of increasing research interest. The citation of articles by physical geographers in *Progress in Physical Geography* reflects the fact that much of the research has been undertaken by other atmospheric scientists but that it is apparent that such research results should be acknowledged by physical geographers and absorbed and utilized in their future research endeavours. The refinement of numerical models and their application to time-bound problems such as the characteristics of glacial and interglacial climates is now offering a link between studies of contemporary climates, of quantitative modelling approaches and of Quaternary environments that the physical geographer cannot afford to ignore.

Considerable progress has been achieved in soil science and such achievements are important to the study of soils within physical geography. Perhaps two groups of achievements may be detected in that there have been advances in technique similar to the investigations of short-term fluctuations of climate, and also there have been conceptual advances analogous to the use of numerical models in the analysis of climates. The criteria for the identification of palaeosols have been refined particularly in relation to the distinction of soil, regolith and weathering profiles and also to the use of laboratory techniques such as using the forms of phosphorus present or the level of amino-acid nitrogen to indicate the biological activity (Gerrard, 1981). Other advances have devolved upon analysis of the humon or decomposition and distribution of organic matter which allows the original population of plants and animals and their associated environments to be interpreted, and the infrared absorption spectra of humic acids can be utilized to recognize palaeosols and to specify the type of former vegetation environment which existed. Further techniques have included the use of weathering indices, the dating of profiles by radiocarbon, the analysis of soil pollen, and of opal phytoliths which are the remnants in the soil profile of the silica originally absorbed by plants and precipitated in plant cells which remain in the soil when the plant decays.

Conceptual advances concerning soil development over time have also been achieved and Birkeland (1974) has provided hypothetical curves for soil development over time under fluctuating and constant climates (Fig. 8.2). In Australia the formulation of the concept of K-cycles (Butler, 1959) provided a methodology which has subsequently been useful in elucidating the stratigraphy of soil development. This included the concept of the groundsurface to represent the development of a soil mantle and the recognition of K-cycles each of which is composed of an unstable phase (K_u) and a stable phase (K_s). In each K-cycle erosion and/or deposition in the unstable phase is succeeded by soil profile development in the succeeding stable phase and in a particular landscape there may be evidence of up to 8 K-cycles preserved in the soil landscape.

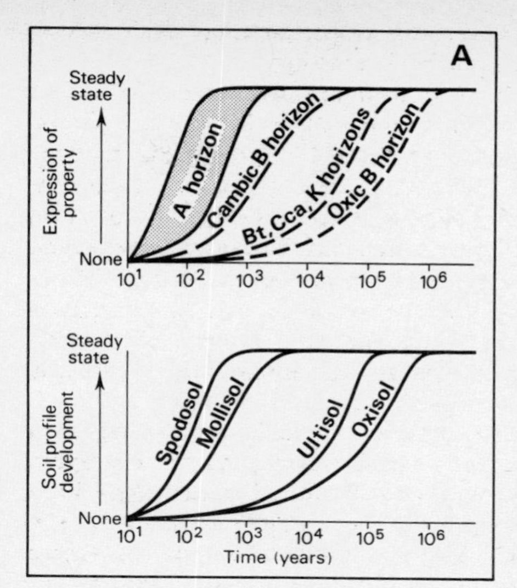

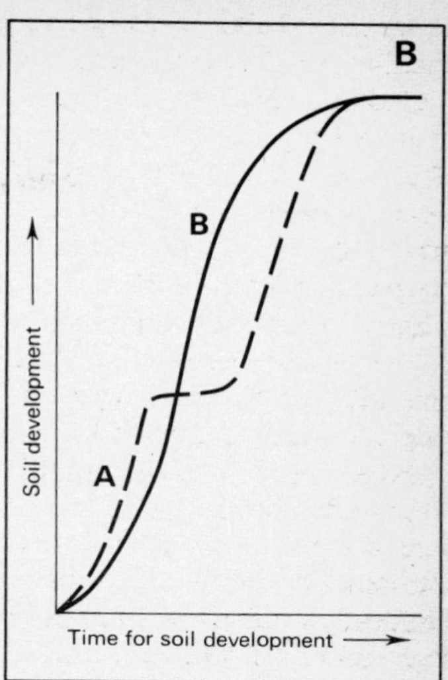

Figure 8.2: Soil variation in time (after Birkeland, 1974). In **A** the time required to attain a steady state is shown for various soil properties (above) and for soil orders (below). The time necessary to achieve a steady state varies with soil property and soil forming factors and hence a soil profile will achieve a steady state only when most of the diagnostic properties are in a steady state.

In **B** hypothetical curves of soil development with time are suggested for areas with a fluctuating climate (A) and a constant climate (B). Variations in B can occur because especially with sandy parent materials development could accelerate because accumulation of organic matter and clay in the soil increases water-holding capacity which promotes clay formation until the steady state is approached.

Only since 1971 has emphasis been placed on the way in which soils as related to landscape develop as a whole and in addition to the K-cycle approach Huggett (1982) refers to the review provided by Daniels, Gamble and Cade (1971) and to the temporal nature of soil–landscape relationships proposed by Vreeken (1973; 1975) to connote the overall topographical influence on soil properties within a drainage basin. When datable land surfaces are found it should be possible to develop chronofunctions which show how soil properties change with time and although Birkeland (1974) and Gerrard (1981) have reviewed the general relationships of soils, weathering and geomorphology there is still a need for the further development of mathematical models. A model for a soil catena or two-dimensional landscape slice, and for a soil landscape or three-dimensional landscape segment were built by Huggett (1976a) to embrace the redistribution of mobile soil constituents by throughflow. Huggett (1982) concludes that:

> Perhaps the biggest untapped area of research in soil geography is the building of mathematical models which link soil and vegetation processes to geomorphological processes on slopes and in landscapes.

There are several directions in which further progress may be made but one approach to the spatial analysis of soil has been based upon fractal concepts (Nortcliff, 1984). The term fractal refers to temporal or spatial phenomena that are continuous but not differentiable and exhibit partial correlation over many scales, and soil properties are visualized as fractals (Burrough, 1983) because increasing the scale of mapping reveals more and more detail.

In biogeography the investigation of temporal change has been long established (p. 81) although the investigation of palaeoenvironments and the study of land-use ecology have been identified as two major contemporary growth points (Taylor, 1984). Palaeoenvironmental investigation has been influenced by contributions in, and publications from, other disciplines and *Quaternary Palaeoecology* (Birks and Birks, 1980) and *Biology and Quaternary Environments* (Walker and Guppy, 1978) are examples that are very significant for physical geographers.

Two other recent developments have emerged, one concerning the significance of temperate landscapes and the other registering the impact of what Stoddart (1983) has characterized as the 'new biogeography'. Much work in biogeographical and ecological theory has been founded in the temperate and boreal regions of the world and hence the tendency has been for the simple and the geologically young to become accepted as normal. It has therefore been argued that the tropics should be viewed as the norm for developing theory and basic principles (Whitmore, Flenley and Harris, 1982) rather than relying too heavily upon regions where soils and vegetation are generally less than 10,000 years old. During the session from which these papers derived Dr J.R. Flenley suggested that the striking degree of ecological change in the tropics during the Pleistocene, as indicated by records of glaciation, of dune fields, of higher lake levels, of ocean bed sediments, and of evidence from fossil pollen, makes it easier to view the tropics rather than the temperate regions as the norm in biogeography. Such a view is further supported because both in temperate as well as in tropical regions: 'the Pleistocene, rather than the Holocene, is the norm in ecology' (Flenley, 1982 in Whitmore, Flenley and Harris, 1982).

Perhaps potentially even more significant is the advent of a new approach in biogeography although Stoddart (1983) notes that a number of recent books have not identified 'the swirling controversies which have dominated the research frontier of the subject for a decade.' The new approach (e.g. Nelson and Platnick, 1981) is founded upon Hennig's phylogenetic systems usually termed cladistics, in contrast to approaches based upon conventional evolutionary taxonomy. Cladistics as a method of classification utilizes graphs of relative affinity termed cladograms which do not require any *a priori* assumptions about the nature of the relationships, including the evolutionary relationships, which are involved. Such a cladistic approach has led to rejection of the methods employed by evolutionary taxonomists and then to different degrees of denunciation of the Darwinian contribution (Stoddart, 1983). When considering the processes that are responsible for distributional patterns in space and time it has been suggested (Croizat, 1978) that these arise from either chance dispersal or vicariance. Dispersal has come to be associated with Darwinian biogeography whereas vicariance biogeography (Nelson and Rosen, 1981) is concerned with biotic distributions which may be congruent with plate-tectonic reconstructions although there are other explanations for vicariant distributions including Pleistocene climatic changes and sea level changes (Stoddart, 1981).

The literature applying cladistic methods to a number of biogeographic problems has been reviewed by Stoddart (1983), who has suggested (Stoddart, 1978; 1983) that both vicariance biogeography and dispersal are appropriate tools at particular levels of enquiry, and that only Udvardy (1981) has attempted to specify the scales which might be appropriate as a background for research investigations. In distinguishing three scales Udvardy (1981) has offered a scheme which is very similar to that proposed in geomorphology by Schumm and Lichty (1965). The three scales that were suggested are the secular scale with spatial dimensions of about 100 km and time dimensions of about 100 years; the millenial scale which covers at least post-Pleistocene time and spatial scales of up to 1000 km where climatic and sea level change are the major factors operating; and evolutionary time or the phylogenetic scale where the time-scale may be up to 500 million years and the spatial extent may reach 40,000 km so that continental displacement may be important at this scale. It is to be hoped that such a tripartite scheme may afford the basis for reconciliation of divergent views in biogeography in the way that the Schumm and Lichty (1965) approach facilitated reconciliation of opposed viewpoints in geomorphology.

In geomorphology there have been advances catalysed by the advent of a range of new techniques. Such techniques have greatly increased the quantity and quality of environmental information which could be employed to reconstruct geomorphological changes, and they have been particularly effective in facilitating use of data derived from studies of processes and of sediments and deposits. Studies of processes have allowed the refinement of rates of erosion and such rates of denudation are important in understanding the way in which change takes place over time. In his review of 'the everlasting hills' Linton (1957) had shown how knowledge of contemporary rates of erosion could be utilized to indicate that the production of a planation surface by subaerial erosion would take of the order of 10 to 110 million years. Subsequently Schumm (1963a) combined knowledge of rates of denudation with information on rates of tectonic uplift and was able to compare rates of denudation with rates of orogeny. Although depending to a considerable degree on location and size of area studied this important paper proposed that 1.0 m per 1000 years is an average maximum of denudation and that this is an order of magnitude less than the average rate of orogeny which approaches 8.0 m per 1000 years.

Advances in dating techniques (Table 4.2, p. 72) have contributed significantly but perhaps the most significant advance for environmental reconstruction has been the use of scanning electron microscopy. Since the 1960s it has been possible to examine the surface of sand grains in detail (Krinsley and Doornkamp, 1973) and subsequently other materials studied under the electron microscope have included organic materials, tills and soils (Whalley, 1978). Further developments including cathodoluminescence and high-voltage electron microscopy furnish detail that even the normal electron microscopes cannot provide (Bull, 1981). The technique has now been utilized for several kinds of environmental reconstruction and was used by Derbyshire (1983) to demonstrate the degree of the cementation and the development of overgrowths in the loess deposits of central China; and from six colluvium sites in Swaziland Goudie and Bull (1984) were able to show that marked edge abrasion of the quartz grains in the uppermost beds did not occur in the lower beds, so that changes in slope process during deposition of the colluvium were inferred and ascribed to greater surface roughness due to exhumation of more core stones and of differentially weathered rock.

In addition to the utilization of such new techniques there has been considerable progress based upon, or at least stimulated by, investigations of contemporary processes. Perhaps most significant have been advances stimulated in the field of fluvial geomorphology but also having relevance to hydrology and to the development of slopes. In a very significant paper in 1965 Schumm provided a basis for the understanding of Quaternary palaeohydrology. Using three relationships, relating mean annual precipitation to mean annual runoff, to mean annual sediment yield, and to mean annual sediment concentration all for different values of mean annual temperature, Schumm (1965) was able to provide a basis from which possible hydrological and geomorphological changes could be interpreted. Subsequently river metamorphosis was proposed by Schumm (1969) as describing those changes of river channels which could be instigated as a result of changes in discharge and of sediment load, and such changes were visualized against the background of a classification of alluvial channels (Schumm, 1963b). A number of studies of specific river channels and drainage basins led to the investigation of relationships between changes of river channel shape and platform as responding to changes of discharge and of sediment transport. Such studies allowed Schumm to propose relationships between the parameters involved so that it was possible to indicate in a particular location how river channel adjustments could take place. Much of the research undertaken by Schumm has now become classic and is summarized in his book (Schumm, 1977), in which he adopts a structure for the idealized fluvial system embracing the zone of production, the drainage basin (zone 1); the zone (2) of transfer; and the zone (3) of deposition.

It is difficult to do justice in a few words to the very substantial contribution made by Schumm but perhaps the contribution made and the subsequent studies which have been inspired can be summarized in four ways. First, it is important to remember that the contribution was developed at a time when other important approaches were made to the fluvial system. These included work on underfit streams by Dury which led to publications including three Professional Papers (Dury, 1964a, 1964b, 1965) which explored the implications of stream shrinkage due to climatic change, and also the work of Wolman, who indicated in a well-cited paper (Wolman, 1967a) how urban stream channels contrast with rural ones and in a less well-known paper (Wolman, 1967b) explored the characteristics of river channel adjustments downstream of reservoirs as well as downstream from urban areas.

At the end of the 1960s there was, therefore, interest in palaeohydrology and river channel changes and subsequent research has been able to demonstrate, secondly, the numerous ways in which change can occur following flow regulation due to reservoirs, urbanization or other land use or channel changes, to document the magnitude of river channel cross-section and channel planform change, and to proceed towards identification of the reasons for change (e.g. Gregory, 1977). Consequently there has been a great advance in the ways in which river channel adjustments can be anticipated (Gregory, 1981) and the degrees of freedom of the fluvial system are being clarified (Hey, 1978), although there is still considerable uncertainty about the exact way in which a river channel will adjust in a particular area (Burkham, 1981).

A third development has been to illuminate river channel changes and drainage basin adjustments by experimentation; at the University of Colorado Schumm employed a 9 × 15 m rainfall erosion facility to investigate drainage network change and channel adjustments (Schumm, 1977). Fourth, and perhaps most significant, has been the way in which concepts of more general application have emerged from these research investigations and this has certainly enhanced understanding of environmental change. In particular the concepts of geomorphic thresholds, of complex response and of episodic erosion, have enhanced the understanding of environmental change. Thresholds were defined by Schumm (1979) to include extrinsic ones which are the levels at which a system responds to an external influence such as climatic change; intrinsic thresholds which are crossed for example when an internal variable changes in the way that long-term weathering reduces the strength of slope materials until slope failure may occur; and geomorphic thresholds which are inherent in the manner of landform change and are thresholds of landform stability that is exceeded either by intrinsic change of the landform itself or by a progressive change of an external variable (Schumm, 1979).

The concept of geomorphic thresholds leads away from the previously accepted ideas of progressive erosion and progressive response to altered conditions and instead the fluvial system seeks a new equilibrium by complex response, whereby the way in which new equilibrium will be achieved will vary from one area to another. Major changes may be accomplished by episodes of erosion and deposition that are the basis of what Schumm has termed episodic erosion. These ideas were incorporated into a modified concept of the geomorphic cycle which accommodates the notion of episodic erosion by dynamic metastable equilibrium in the earlier stages of the cycle (Schumm, 1979).

Thresholds are invaluable in the understanding of temporal change because they should specify the process boundary conditions at which change, which may be expressed in morphological adjustments, will occur. Therefore in the development of studies of temporal change there has been an increasingly enthusiastic search for thresholds, although it has been realized that such threshold conditions are very difficult to specify in simple terms. One very good example is the distinction between gullied and ungullied valley floors in the Piceance Creek area, Colorado, where Patton and Schumm (1975) plotted a linear relation between valley slope and drainage area as the threshold separating the gullied from the ungullied valley floors. Subsequently this type of approach has been further developed by W.L. Graf now of Arizona State University. He studied the Central City district of Colorado where mining had developed rapidly in the nineteenth century and subsequently declined just as rapidly. He was able to establish (Graf, 1979b) a relationship between tractive force in dynes calculated for a 10-year recurrence interval and biomass on the valley floor in $kg.m^{-2}$ which provided a threshold between the entrenched and the uncut valley floors. This extremely stimulating paper provided a means of expressing the relation between force and resistance in process terms and identified a threshold situation that could be employed to interpret the spatial distribution of valley floors entrenched since the mining activities of the 1830s; such a relationship can be employed to indicate areas of valley floor instability at present and in the future.

Such developments potentially have great significance in geomorphology where Graf has argued (1979b, p. 266) that 'the tradeoff between force and resistance lies at the heart of explanation in geomorphology.' However, threshold relationships have often been regarded as linear ones whereas the incidence of change may be rather more complicated particularly because of the Hurst phenomenon whereby there is a tendency for a non-periodic grouping of similar values over long periods of time and 'a short-term realization of the event sequence in nature will not sample all scales of variability in the process and that, in consequence, distribution statistics derived from the sample realization will be biased' (Church, 1980). If the relation between an input series (X_t), or forcing function as it is sometimes called, and the output series (Y_t) is envisaged as a transfer function $Y_t = g X_t$ where g is the impulse response function, then it has often been assumed that the threshold relation will be linear. However, several complications may occur (Church, 1980) because the history of the landscape may dictate that a specific situation is supply limited, and that the output sequence may result from the combined effect of several conditions rather than a single one. Therefore the outcome of this pattern of complex response is that geomorphological event sequences may be more intermittent than the forcing sequence (Church, 1980, p. 18).

It has therefore been necessary to extend consideration of thresholds in at least two directions, the specific and the general. Specifically has been the investigation of the significance of large rare events to establish their significance (Starkel, 1976) and this has shown that in some cases it is necessary to envisage a change in boundary conditions for extreme events because very high sediment and water discharges may behave as debris flows as Jarrett and Costa (1983) have shown in the case of the foothill streams in Colorado. There may be a difference in the significance of events between temperate and arid areas and Wolman and Gerson (1978) have suggested that the absolute magnitudes of climatic events and absolute time intervals between the events are not in themselves satisfactory measures of the geomorphic effectiveness of events of different magnitudes and recurrence intervals. They proceed to propose a time-scale for effectiveness that relates recurrence intervals of an event to the time needed for a landform to recover to the form that existed prior to the event occurring. Measured recovery times quoted in the literature range from less than a decade for some tropical regions to decades or more in temperate regions, and recurrence intervals of high-magnitude storms which trigger mass wasting range from 1 to 2 years in some tropical areas, to 3 or 4 per 100 years in some seasonal rainfall areas, and to 100 or more years in some temperate regions (Wolman and Gerson, 1978).

A rather more general approach to thresholds has emerged by considering the extent to which parameters involved in disturbed environmental systems correspond in the timing of their disruption and then in considering the temporal sequence of disruption. The timing of disruption was considered by Knox (1972) when studying valley alluviation in south-west Wisconsin, where he found a disunity between hillslope processes and channel processes. Thus during shifts from humid to arid climatic regions (Fig. 8.3A) there may be time lags in the rate at which the relative vegetation cover (B) responds and the way in which hillslope potential for fluvial erosion (C) and geomorphic work change. Douglas (1980) has suggested that such lags have to be removed when contemporary process–response systems are quantified and that the possible existence of lags and their

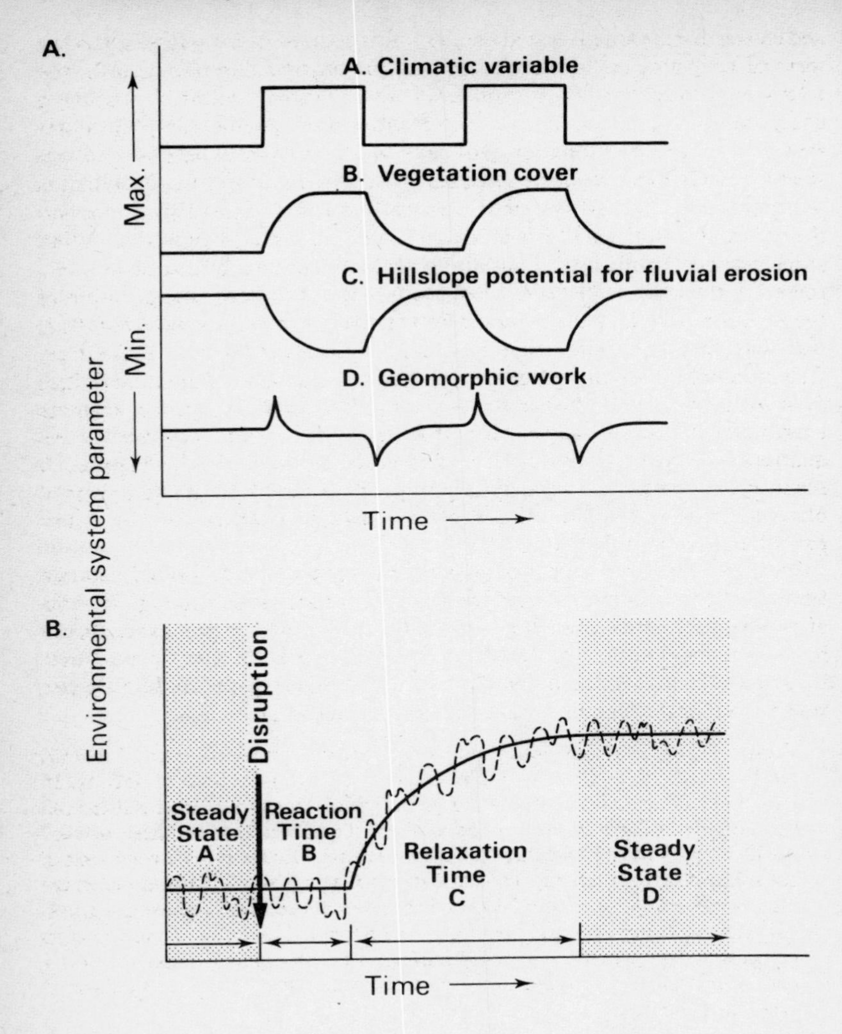

Figure 8.3: Schemes of geomorphological change. **A** suggests how environmental parameters are phased when responding to disruption (after Knox, 1972). **B** portrays the response of a geomorphic system to disruption (after Graf, 1977).

magnitude in past environments should be investigated.

An intriguing approach to the timing of disruption was provided by applying the rate law (Graf, 1977) in geomorphology. Because the impact of human activity is primarily expressed in disruptions of natural processes and in the disturbance of steady states, Graf proposed that a rate law in the form of a negative exponential function, similar to that used to describe the relaxation times of radioactive materials and chemical mixtures, provides a useful model for relaxation times in geomorphic systems. Graphic representation of the disrupted system (Fig. 8.3B) involved the change from steady state A to steady state D passing through reaction time B, which is the time needed for the system to absorb the impact of the disruption, and the relaxation time C which is the time period during which the system adjusts to new conditions. In addition to the magnitude of disruption, and therefore the difference D–A, it is also important to know the duration of the reaction time and the relaxation time. Using the rate law Graf (1977) was able to demonstrate that if a small fluvial system in the Denver, Colorado area is disrupted either by climatic change or human activities then the system will adjust by eroding a gully towards a new steady-state length and that it will approach half that length in about 17 years, three-quarters in 34 years, seven-eighths in 51 years and so on. Such an approach illustrates how new concepts of time, capable of extension to other parts of physical geography, are affording an exciting prospect for the interpretation of temporal change.

Progress in other branches of geomorphology has also embraced concern for thresholds and has made progress towards new models although the rate of progress in other branches may not have been as rapid as in fluvial hydrogeomorphology simply because there have been fewer researchers involved. Thus in a report on slopes and slope processes in 1980 it was suggested (Mosley and O'Loughlin, 1980, p. 103) that:

> Clearly the literature on slopes and slope processes during the past 12 months has been voluminous, but with few exceptions it contains little of novelty. In fact, much of the work is depressingly unoriginal, the work on surficial erosion in particular following closely the research being carried out three or four decades ago. The most worthwhile advances currently seem to be occurring in the subfield of mass wasting and slope stability, where physical geographers are increasingly adopting methodologies from other disciplines. The trend towards convergence brings to mind once again the perennial question of the reality of physical geography as a separate discipline; its supposedly distinctive subject matter, methods and approaches seem to be increasingly replaced by those of other disciplines.

Two years later (Mosley, 1982) it was concluded that use of methods from other disciplines had certainly been beneficial, but was being succeeded by indications that geomorphologists were beginning to see beyond the processes to the landscape, so that with the expertise in process studies achieved one may expect significant advances in the application of process results to studies of slope and landscape evolution. In fact research experience of particular branches of geomorphology may explain why some see the need for parallel studies of past and of present processes (Starkel, 1982) whereas others (Douglas, 1982) contend that the contribution of studies of earth-surface processes to landform evolution is an unfulfilled promise. In studies of arid areas by collating large amounts of information on contemporary processes it is possible to illuminate the past, and, for example, Goudie (1983) has collected details of the environmental consequences and the global distribution of dust storms and then proceeded to indicate the way in which dust storms have varied through time.

The application of knowledge of contemporary environmental processes to problems of past landscape development is very often dependent upon the ergodic hypothesis whereby space has been substituted for time. An illustration of how effective this can be is provided by the schematic representation of alas development during the development of thermokarst by degradation of permafrost (Czudek and Demek, 1970). By analysing the spatial distribution of alas character they were able to provide a 6-stage model showing the progress of a surface with syngenetic ice wedges to a thermokarst valley. Other examples occur in one of the publications arising from Binghamton symposia devoted to space and time in geomorphology (Thorn, 1982).

In the field of glacial geomorphology notable developments have also occurred. Although very interrelated to work by scientists in other disciplines and characterized by research progress organized in subfields including basal ice and glacier beds, cirques, glaciohydrology, there have been significant attempts to model climates, glaciers and ice sheets. Thus Sugden (1978) provided a reconstruction of glacial erosion by the Laurentide ice sheet and concluded that landscapes of glacial erosion are equilibrium forms fashioned largely when ice sheets are at a maximum and reflect periods of steady-state conditions of the order of 100,000 years.

After constructing a steady-state model of the late Devensian ice sheet which covered much of Britain (Boulton, Jones, Clayton and Kenning, 1977), patterns of glacial erosion and deposition over Britain were summarized and it was suggested that the central parts of the ice sheet were relatively inactive, that intense glacial erosion did not occur in highland Britain during glacial maxima and that relatively high erosional intensities in marginal areas were produced by high marginal velocities which in turn precluded thick lodgement till and drumlin formation near the margin and concentrated them in internal zones. At least two themes characterize recent research in the field of glacial geomorphology therefore. First has been the investigation of contemporary processes and the development of time sequences as illustrated by glaciohydrology where the meltwater sources and short-term temporal variations are being explored. Secondly are studies such as those by Sugden (1978) and by Boulton, Jones, Clayton and Kennedy (1977) where the pattern of deposits and features is related to, and interpreted in the light of, a model sequence of ice sheet development.

More general progress

A number of themes emerge from consideration of developments in studies of temporal change in the branches of physical geography and in particular several strands are now common to several branches of physical geography and interdisciplinary interaction is becoming more evident. More general progress is often achieved when the model produced in one branch of research is capable of application elsewhere and for example the rate law (Fig. 8.3B)is capable of application in branches of geomorphology other than fluvial, and the notion of lags in geomorphic systems (Fig. 8.3A) finds analogies elsewhere as for example in glacier response to climatic shifts as shown by Goudie (1977). In palynology too there is a movement beyond the investigation of site history and Edwards (1983) suggests that there are already sufficient data available to resolve many problems, albeit at a different level of generalization, and he suggests (Edwards, 1983, p. 120) that:

> Quaternary palynology . . . as an academic discipline would surely be aided by a little less microscope work, more use of available data and, optimistically, the published overviews of specialists concerning their own chosen area of palynology.

Research in palaeoclimate mapping from pollen data and on the vegetational response to climatic change has involved developing transfer functions to relate pollen spectra and climate variables and migrational lags which are evident as vegetation changes in response to climatic changes (Fig. 8.3) but not necessarily synchronously (Prentice, 1983). One way of furthering this research (Prentice, 1983, p. 281) would be:

> to build an equilibrium model, deriving optimum-surface response functions from modern climate and pollen or tree abundance/distribution data, and then attempting to reconstruct Holocene migrations and retreats under various plausible climatic scenarios. Residuals from the simulation might give an idea of the extent of migrational lag. Such techniques are potentially far more eloquent than verbal arguments.

Thresholds have now been identified throughout geomorphology and Coates and Vitek (1980) note that although the ideas of critical limits, boundary conditions and yield points form important features in other disciplines, such terminology formed a comparatively small part of the geomorphological literature. They suggest that the two doctrines of catastrophism and uniformitarianism should be supplemented by a third base which is concerned with thresholds. In their edited book they group the papers as concerned with historical background, with fluvial landforms, with hydrogeologic regions, with thresholds in other geomorphic processes, and finally with thresholds and man. The spectrum of application is now very considerable throughout geomorphology and thresholds defined as (Fairbridge, 1980, p. 48).

> . . . a turning point or boundary condition that separates two distinct phases of interconnected processes, a dynamic system that is powered by the same energy source,

may be applied to studies of tectonics, hydrological processes, glaciology, eustasy and sedimentology. Glacier surging that is characteristic of the flow of some glaciers exemplifies the threshold situation because once the threshold is exceeded the surge occurs with velocities often several orders of magnitude larger than normal and the glacier then becomes quiescent with the lower reaches raised in elevation and the upper reaches lowered. Subsequently instability is built up again as the upper part rises and the lower part ablates until another surge occurs (King, 1980b).

Thresholds therefore relate to energy transfers that are increasingly the focus of attention arising from a systems viewpoint (chapter 7, p. 153) and there is no doubt that increasing focus upon the way in which thresholds provide the boundary conditions for phases of energy transfer will prove to be a very promising basis for further developments in physical geography. In investigating the spatial variation of fluvial processes in semi-arid areas Graf (1982) investigated the controls upon spatial variation of fluvial processes, the way in which energy varies spatially in channel networks, and the geomorphic consequence of the spatial variation of fluvial processes. In a subsequent paper on the Henry Mountains, Utah he was able to suggest the way in which stream power varied along the channel in 1883 which contrasted with the downstream increase in power that obtained in 1909 and 1980 (Graf, 1983).

A major difficulty in dealing with threshold situations is that many analyses undertaken in physical geography have utilized linear relationships whereas a threshold automatically involves a different form of distribution. Linear relations usually established by regression have been the subject of improper use in the earth sciences as shown by Williams (1983), and such relations have been improved upon in at least two ways. Firstly, there are step functions which provide a method of resolving a series into distinct sequences. Thus Dury (1980b) analysed precipitation data from Sydney and subsequently (Dury, 1982) used step functions in the analyses of long streamflow records. The method involves reduction of a time series to stationary form by calculation of the series mean and then the calculation of cumulative deviations from the mean. If a graph is plotted of cumulative deviations against time it is then possible to differentiate blocks of years in the records. Where thresholds occur influencing a sequence then this should be indicated by the step function sequence.

A second approach which may be capable of wider application is provided by catastrophe theory, which was proposed by Thom (1975) for the description and prediction of a number of discontinuous processes. This can therefore relate to changes which are catastrophic transitions or flips of the system state from one domain of operation into another perhaps irreversible one (Chorley and Kennedy, 1971). Seven elementary catastrophes were provided (Thom, 1975) and of these the cusp catastrophe may be utilized for modelling temporal changes in environmental systems. The original theory was developed by Thom (1975) to describe temporal development and/or evolution where the growth or development of an organism occurs as a series of gradual changes inspired by, and giving, catastrophic jumps which are linked to large-scale changes in the organism (Stewart, 1975; Bennett and Chorley, 1978). Graf (1979a) has reviewed the pertinence of catastrophe theory in fluvial geomorphology as being useful for describing the behaviour of geomorphic systems although it does not answer the basic question 'why?'. The cusp catastrophe is characterized by abrupt and smooth changes, by divergent and bimodal behaviour, hysteresis, and stability of structure. Graf (1979a) perceived the advantages to arise from the combination of concepts of equilibrium and change, stability of the change structure and the perspective provided which is unlike earlier models, whereas the disadvantages arise from the difficulty of identifying system control factors, of defining energy functions and from the generality of the theory. The cusp catastrophe has been used (Thornes, 1980) to model the behaviour of ephemeral channels in terms of spatial and temporal variations in sediment transport and channel width. Further applications in physical geography are possible but initial enthusiasm may reflect what Alexander (1979) termed 'the geographer's insatiable appetite for new methods of analysis irrespective of the constraints of data'.

Particularly in geomorphology these approaches to temporal change have been taken a step further. Considering landscape sensitivity and change Brunsden and Thornes (1979) derived a series of four fundamental propositions of landform genesis generated by process-form studies and these were:

Constant process – characteristic form – For any given set of environmental conditions, through the operation of a constant set of processes, there will be a tendency over time to produce a characteristic set of landforms.

Transient behaviour – Geomorphological systems are continually subject to perturbations, which may arise from changes in the environmental conditions of the system or from structural instabilities within. These may or may not lead to a marked unsteadiness or transient behaviour of the system over a period of 10^2 – 10^5 years.

Complex response – The response to perturbing displacement away from equilibrium is likely to be temporally and spatially complex and may lead to a considerable diversity of landform.

Sensitivity to change – Landscape stability is a function of the temporal and spatial distributions of the resisting and disturbing forces and may be described by the landscape change safety factor, considered to be the ratio of the magnitude of barriers to change to the magnitude of the disturbing forces. They also suggested a transient-form ratio (TF_r) which can express the sensitivity of landform to both internally and externally generated changes in the form:

$$TF_r = \frac{\text{mean relaxation time}}{\text{mean recurrence time of events}}$$

It was argued (Brunsden and Thornes, 1979) that the investigation of long-term change should involve establishing the characteristic response to fixed distributions of barriers to, and forces of, change using calibrations from studies of contemporary processes, and that these could be accompanied by evaluation of the relative sensitivity of parts of the environmental system.

Further need for theory in geomorphology has been stressed (Thornes, 1983b) in recognizing that the two frameworks evident in geomorphology since 1950, devoted respectively to chronology (chapter 4) and to processes (chapter 5), will only be reconciled when a model of long-term geomorphological behaviour is developed. An approach towards evolutionary geomorphology is provided (Thornes, 1983b) by visualizing process domains which may overlap, correspond, compete with or be exclusive of, other process domains. The concept is analogous to the niche in ecology, and domains represent the equilibrium relationships between processes according to the controlling parameters. Whereas process geomorphology is concerned mainly with the way in which domains are determined, evolutionary geomorphology is concerned with the initiation and development of the structure giving rise to the domains. It is possible to envisage changes in the relative importance of different processes and in addition a particular process may change its domain and interact with that of another process. To analyse development of this kind may require consideration of unstable behaviour which has been pursued as an alternative to the steady-state model. In an unstable system there is no return to equilibrium and it may not be possible to predict successive behaviour. A theoretical treatment of the conditions under which a hillslope hollow or rill will either fill with sediment or develop to form first a channel and then a valley (Smith and Bretherton, 1972) introduced the possible emphasis upon growth and change rather than upon dynamic equilibrium and this has been developed by Luke (1974) and by Kirkby (1980). Thornes (1983b) concludes that emphasis in geomorphology is shifting from the observation of equilibrium states *per se* to the recognition of the existence of multiple stable and unstable equilibria and this may be the basis for new models of geomorphological evolution.

In the context of human impact on environment Oldfield (1983) has argued that physical geographers have hitherto emphasized models of ecosystem change which are deterministic, progressive and evolutionary in character. The succession–climax model is still very evident in biogeography despite its weakness in the light of recent empirical evidence and the difficulties encountered when trying to reconcile it with systems theory. In addition, models with a cyclic or harmonic element have been developed in view of variations reconstructed at a wide variety of scales. However, Oldfield (1983b) argues that there is a need for a model of ecological change which differs in principle from both the evolutionary and the cyclic models and is more analogous to the steady-state models developed in other branches of science. He proposes that lakes together with their sediments and drainage basins provide an appropriate focus which can utilize the new methods available (Oldfield, Battarbee and Dearing, 1983) for analysing and dating the flux of materials. The steady-state model proposed (Oldfield, 1083b) is in many ways analogous to models employed in geomorphology (as in Figure 8.3B) and extensive shifts between steady states have often been triggered by processes of soil depletion and erosion with the associated consequences for soil structure, water content and nutrients.

Interdisciplinary awareness

The fear that greater specialization inexorably leads towards fissiparist tendencies increasing the divisions between branches of physical geography has not been substantiated by recent trends for two groups of reasons. Firstly because of the similar models and conceptual approaches which are common to branches of physical geography and secondly, because of the greater links with other disciplines.

The first trend has been evidenced by several examples in the last few pages. The way in which biogeographic studies have progressed towards the lake–drainage basin system necessarily fosters links between biogeography, geomorphology and hydrology and the ecosystem watershed concept has been advocated (O'Sullivan, 1979). In environmental systems there is a recurrent interest in domains by the geomorphologist which is reminiscent of the niche used by the ecologist, and the need to progress towards a vision of time scales is as fundamental in geomorphology (Schumm and Lichty, 1965) as it is in ecology (Udvardy, 1981). Investigation of temporal change and of alterations in energy transfers has demonstrated that markedly episodic influences can be very significant in geomorphological sequences and also in soils as embraced in the K-cycle concept, and thresholds are also the focus of attention in several branches of physical geography.

It is notable that the trends which recur through several branches of physical geography are promising greater coherence of the branches of the subject at a time when very stimulating approaches are emerging. However a second trend is towards greater liaison between the several disciplines concerned with environmental change. Thus some of the themes noted above as characteristic of several branches of physical geography have also been prominent in, and sometimes have been imported from, other earth sciences. Thus thresholds have achieved prominence in ecology (May,

1977) and there is further potential for employing fractals in relation to environmental data. Advances in geostatistical studies should furnish the basis for greater appreciation of the products of temporal change. Fractal dimensions, which range from 1, which is completely differentiable, to 2, which is so rough and irregular that it effectively takes up the whole of a two-dimensional topological space, have now been estimated for various environmental series by Burrough (1981) and this has implications for the spatial analysis of soil properties (Nortcliff, 1984). Interdisciplinary awareness has also occurred as there has been increasing cooperation between scientists in different disciplines. Many examples of recent research investigations could be cited to illustrate this point but fertile cooperation between a physical geographer and a geologist has enabled elucidation of the way in which the sedimentary infill relates to the generation of discharge and shows how this can assist palaeoenvironmental interpretation (Frostick and Reid, 1977). Similarly Prentice (1983) has shown how research on postglacial climatic change is benefitted by contributions from several disciplines because 'Cross-checks among paleoclimatic estimates obtained in different ways will be extremely useful, but no one method or discipline is privileged.' Fluctuating boundary conditions will cause systems to respond in ways which vary according to the characteristic response times and to the frequency spectrum of the fluctuations (Prentice, 1983) and such problems are best elucidated using resources and expertise from several disciplines.

Interdisciplinary cooperation of this kind has often happened incidentally but there are cases where deliberate cooperation has been fostered. Thus the International Geological Correlation Programme embraces a number of interdisciplinary and international research investigations which involve physical geography researchers. Some of the investigations are listed in Table 8.2 and many of these involve contributions from a range of disciplines. Project 158 is designed to reconstruct the palaeohydrology of the temperate zone during the last 15,000 years and is organized in two subsections, one devoted to fluvial reconstruction (158A) and one to the analysis of information from lakes and mires (158B), and the general objectives have been summarized by Starkel (1983) and by Berglund (1983). Such research has demanded the use of comparable methods in each of the participating countries, it has involved contributions from geologists, ecologists, archaeologists, hydrologists as well as from physical geographers, and it is already providing manuals of techniques (e.g. Starkel and Thornes, 1981) progress reports and integrated studies (Gregory, 1983; Starkel, 1981). Such progress is already furnishing a greater understanding of environmental change. Such studies of environmental change are not restricted to the past but may also have implications for the future. Thus desertification is one of the problems which relates to past environmental change and is also inextricably linked with future changes. In a review of climate on the desert fringe Hare (1983) concluded that:

Table 8.2: IGCP (International Geological Correlation Programme) Projects relevant to physical geography

Project number	Title	Duration of project
24	Quaternary glaciations in the northern hemisphere	1974–83
61	Sea-level movements during the last deglacial hemicycle	1974–82
129	Laterization processes	1975–83
143	Remote sensing and mineral exploration	1976–82
146	River flood and lake-level changes	1976–84
158	Palaeohydrology of the temperate zone	1977–87
184	Palaeohydrology of low latitude deserts	1981–85
200	Sea-level correlation and applications	1983–87
201	Quaternary of South America	1983–87

As listed in *Geological Correlation* II, May 1983 Paris: Report of the International Geological Correlation Programme (IGCP), Unesco.

> Climatologists, geomorphologists and ecologists work alongside archaeologists, engineers and physicians to try to get a good, interdisciplinary view of the problem. It is characteristic of the major geographical margins of the world – the desert fringe and the Arctic and alpine transitions – that they call for such broad approaches. . . . Truly satisfying linkages need actual hypotheses as to the physical and biotic processes at work. Current climatic geomorphology trends in this direction, though past work has depended heavily on spatial correlations. Until these processes of interaction are formally quantified, we are still, to repeat a disparaging metaphor, engaged in a hobby like bird-nesting or train spotting.
>
> But the way is open for a sounder approach in linking climatology with ecology and geomorphology.

Such a sounder approach may be the basis for illuminating future as well as past temporal change. This link between elucidation of the past and estimation of the future was exemplified by a review of climate, drought and desertification (UNESCO, 1984) based largely upon the contribution of Professor Hare, by chairing an expert group meeting convened by the World Meteorological Organization, and producing a general paper for the Agricultural Meteorology Commission of the World Meteorological Organization. Because it is not easy to separate applications of physical geography from the investigation of temporal change the next chapter should be seen particularly in relation to the content of the last three chapters.

9

Advancing applications

Physical geographers have shown a considerable reticence in becoming involved in applications of their research. This may have occurred because the prevailing ethos has always been to direct attention towards all the factors influencing a particular situation, which has led to a reluctance to propose a single solution for a particular problem. Such innate reticence may have arisen because physical geographers were studying inappropriate subjects or were working at the wrong scales. Subjects may have been inappropriate because an initial emphasis upon historical development and upon form and description was succeeded by emphasis upon process and upon advances in chronology, and only when these approaches were more completely integrated, as indicated in chapter 8, could physical geography be in a position to contribute more definitively to applied questions. Indeed, applied problems usually depend upon questions of temporal or spatial extrapolation and so it is quite logical that applications should arise from, and interact with, studies of temporal change. The wrong scale may have featured in studies hitherto because the tendency has been to ignore the macro-scale and to focus instead upon the micro-scale and meso-scale approaches. Such a focus upon small areas was inevitable when advances were being made in the study of processes, but cannot be sustained indefinitely when some of the major world problems are of global proportions, such as CO_2, desertification, and deforestation. Neglect of the study of the natural environment at a time when the interest of other disciplines was increasing (Manners and Mikesell, 1974) was not simply a reflection of the fact that environmental concern was often focused on a scale that was not coincident with the scale of physical geography research, but also the fact that fissiparist tendencies in physical geography had necessarily made it more difficult to visualize the unity of the physical environment. Therefore in the USSR Gvodetskiy, Gerenchuk, Isachenko and Preobrazhenskiy (1971) regretted this separation into geomorphology, climatology, pedology and biogeography as indicated in the quotation in Table 2.3 (p. 39) and in other papers there have been pleas to remember the totality of the physical environment (e.g. Walton, 1968).

Many instances have been apparent in previous chapters of ways in which applications of physical geography research may arise. Studies of processes and of process–response systems have provided much needed knowledge which has contributed to decision making. Thus research on the solutes and pollutants reaching urban stormwater drainage allowed Revitt and Ellis (1980) to make recommendations about the frequency and nature of street cleaning practices. Similarly establishing the impact of human activity, upon soil salinity for example, could influence the way in which irrigation schemes should be implemented. Although an increasing number of possible applications arise from research on the themes summarized in chapters 4 to 9, there are also some branches of physical geography that have been envisaged as primarily applied and the adjective applied has been placed in front of geomorphology, climatology and pedology to create recent book titles (Hails, 1977; Smith, 1975; Hobbs, 1980; Oliver, 1973; Craig and Craft, 1982; Verstappen, 1983). In this chapter the aim is to indicate how such applied approaches developed; to review landscape ecology and mapping; environmental evaluation and the prediction of change and the design of environment because these themes have hitherto provided the major applications.

Development of applied physical geography

It is perhaps first necessary to define what applications of physical geography really involve and at least three components may be envisaged. Firstly, relevance has been achieved from existing research and many subjects studied by the traditional subdivisions have proceeded towards applicable research. In this way studies of environmental hazards have provided information pertinent to decision making and the drought in Britain 1975–76 was the subject for an atlas which portrayed the physical characteristics and the physical and economic consequences of the event (Doornkamp, Gregory and Burn, 1980). Secondly, and more directly, has been the applied intention in research and in teaching and this has been shown in the appearance of a final chapter devoted to applications in many books; in the increasing popularity of the applied theme as a major ingredient in inaugural lectures (Embleton, 1982; Stephens, 1980; Chandler, 1970; Douglas, 1979); and in the use of the theme for series of books (e.g., R.J.C. Munton and J. Rees, editors of The Resource Management series, Allen & Unwin; K.J. Gregory, editor of Studies in Physical Geography, Butterworths; and K.M. Clayton and J.H. Johnson, editors of Topics in Applied Geography, Longmans). In addition to these indicators, the new journal *Applied Geography* (1981—) published by Butterworths contains many papers devoted to applied physical geography and in Eastern Europe and the USSR the advantages of applied research have been advocated by many writers. Thus Gerasimov (1968) advocated constructive geography as providing: ‘The theoretical basis and practical recommendations for man’s transformation of the environment for the benefit of society’ and subsequently (Gerasimov, 1984) has traced the way in which constructive geography has been used for a variety of studies of the planned transformation of the natural environment to allow effective use of natural resources. Fifteen issues of a special series *Problems of Constructive Geography*, have been published 1975–1980 by the Institute of Geography of the USSR Academy of Sciences.

Thirdly, and possibly most convincingly has been the involvement of physical geographers in a number of applied directions. Such involvement has been as members of international and often interdisciplinary groups, for example concerned with desertification; in consulting, perhaps in conjunction with engineers, as necessary in the case of site engineering projects; in practise, whereby physical geography graduates obtain posts in the water industry, or in nature conservation for example; or on national committees as exemplified by T.J. Chandler on the Royal Commission for Environmental Pollution (Chandler, 1976). During World War I the description of terrain characteristics of the Flanders battlefield is an example of applied physical geography through terrain evaluation for military purposes (Johnson, 1921). In many examples of these three kinds of applications physical geographers have made contributions to decision making although not necessarily as physical geographers and in many instances the contributions have been advisory rather than mandatory.

Since 1970 many have advocated the need to proceed towards applications of physical geography and a number of these views are collected in Table 9.1. Applications in relation to management have been elucidated by Clark (1978) writing in the context of the coastal environment but capable of extension to other parts of the discipline. He envisages geomorphology as a field which has evolved from a purely systematic study to one which has considerable practical application and he argues that the new role for the geomorphologist spans process systems, public information, remedied

Table 9.1: Opportunities perceived for applications of physical geography

Year	Author
1970	T.J. CHANDLER 'Mankind must learn to make the best use of his environment, conserving its main form and using purposeful modification only where the benefits are clear and the ill effects can be shown to be negligible. These, I would suggest, should be the primary aims of climatic management.'
1972	I. DOUGLAS 'As the behavioural, economic and social processes involve a different range of expertise from the less complex physical, chemical and biotic factors, a dichotomy between physical and human geography has appeared in some geographical institutes. However, the greatest geographical contributions to environmental management have come from the blending together of all these factors to understand the true complexity of man-environment ecological systems.'
1983	D.K.C. JONES '. . . there still exists a range of potential inter-linkages between the natural environmental and socio-economic systems which could provide an *additional* focus for geographical research. . . . The development of a new focus would . . . be beneficial to the whole subject, for the research problems encountered at the internal interface are just as demanding and as pertinent as those at the more attractive external interfaces.'

strategy design, and participation in frameworks for decision making. This leads to the conclusion (Clark, 1978, p. 281) that:

> . . . the aim of applied geomorphology is not to prevent or reduce development and resource use, but rather to optimize that use by reducing both costs and impacts.

Some researchers have concluded that applications of physical geography necessarily involve closer liaison of physical and human geography although Johnston (1983c) is not convinced (see p. 198).

Attempting an integrated spatial perspective is often an important function, and in the case of the drought in Britain 1975–76 the Drought Atlas was able to provide three maps which showed the drought intensity, the drought impact, and the relation of the two as the drought stress (Doornkamp, Gregory and Burn, 1980).

There are two interrelated ways of summarizing the achievements so far. First, it is possible to envisage applications of geomorphology, of climatology, of soil geography and of biogeography although such a fissiparist classification is in danger of omitting some of the needs of applied research. In the case of applied geomorphology in Britain, Jones (1980) has distinguished large-scale or 'areal' investigations usually carried out for land development purposes; medium-scale 'linear' investigations such as road lines which involve greater detail; and very detailed 'local' or 'site-specific' studies that are usually for engineering purposes. In his inaugural lecture Embleton (1982) provided examples of ways in which applications could arise from research on specific aspects of glaciology (Table 9.2). Secondly, it is possible to envisage applications developing in four stages as concerned progressively with description and depiction of environment, with environmental impact, with environmental evaluation and with environmental prediction. These four stages can be recognized in each of the branches of physical geography and will be outlined here.

Description of environment in a relevant way is one of the primary requirements for a more relevant physical geography. Whereas mean annual temperature and mean annual precipitation were the data series long quoted by geographers, the number of frost free days, and the length of the growing season are examples of indices of climate more relevant to agriculture and rural land use. There are many examples of the way in which climate can be described in a relevant way, as illustrated in Barry and Perry (1973). In many branches of physical geography it has been possible to characterize the environment according to use for particular purposes. This is well exemplified by landscape ecology, which is the characterization of environment for land use, and this approach has been illustrated by the land systems approach (see p. 198). Environmental geology offers a way of emphasizing many aspects of environment in relation to management problems and the approach which enumerates all pertinent variables, for example in tabular form (Chandler, Cooke and Douglas, 1976), can be

Table 9.2: Some examples of applications of glaciology proposed by Embleton 1982

	Ice	Meltwater	Snow	Ground ice
Positive utilization	Icebergs for water Ice as a refrig-erant Waste disposal in glaciers Artificial ice islands Drilling from ice shelves	Electric power generation	Water storage in drain basins Skiing and tourism Frost protection	Permafrost as a loading-bearing material
Hazards	Advancing/ surging glaciers Glacier falls (ice avalanches) Crevassing and calving Icebergs as transport hazards Sea-ice hazards: Harbour blocking, structural damage River/lake freezing, ice-floe damage Ice accretion (on ships, aircraft, etc.) Loading limits on ice surfaces, on floating ice, etc. Hazards of sub-glacial mining, glacier excavation	Floods on melt-water rivers Bursting of glacially and sub-glacially impounded lakes Damage by sediment load (e.g. in turbines)	Avalanches Snow cover of transport lines, airfields, urban areas Snow loading on structures Snow accretion on electric wires	Degradation of ground ice – thermokarst Aggradation of ground ice – ground heaving, freezing of water-bearing strata Waste-disposal problems
Environmental controls	On climate – global, local On world sea-level			

of great value in establishing the variables involved and in identifying the basic problems.

Environmental impacts have also been investigated in some detail and emphasis has been placed upon the magnitude of environmental impact so that future estimates may be made. In the field of geomorphology, environmental geomorphology was perceived by Coates (1971, p. 6) to be:

> . . . the practical use of geomorphology for the solution of problems where man wishes to transform landforms or to use and change surficial processes. . . . The goal for geomorphic environmental studies is to minimize topographic distortions and to understand the interrelated processes necessary in restoration or maintenance of the natural balance.

The difficulty of establishing environmental impact is well illustrated by the effect of increasing CO_2 concentrations. In the 1970s the increase was noted but there was no general consensus about the significance which this increase could have. More recently it has been generally agreed that a doubling of the CO_2 in the atmosphere may be responsible for an increase of mean sea surface temperature by 3.0°C ± 0.5°C. General circulation models are now becoming sufficiently refined to allow the consequences of such temperature increase to be modelled but there is considerable scope for further estimations to be made in the light of change in the past. Thus Butzer (1980) suggested how world climates in 2050 could be envisaged in terms of those wetter or drier than present and this was conceived in the light of reconstructions of world climates between 5000 and 8000 years ago.

Under the heading of environmental impact there are many ways in which the magnitude and significance of human activity has now been documented. In the investigation of soil erosion Stocking (1981) has described the way in which the rate of gully development may be related in a multiple regression equation to factors which include precipitation, size of drainage area and height of headcut. It is also necessary to know what environmental impacts can occur in a particular situation and also how large these impacts can be. In the case of river channellization it has been shown how extensive such channellization was in England and Wales (1930–80) (Brookes, Gregory and Dawson, 1983) and it is equally necessary to know how the effects of loading or unloading can influence endogenetic processes, for example, by instigating surface subsidence and/or the increase in frequency of minor earthquakes (Coates, 1980). Thus environmental impact can embrace direct impact on environment as well as impact on environmental processes and both can inspire approaches which are pertinent to applied research and to environmental management.

Evaluation of environment and environmental processes is the stage at which research has attempted to show how certain characteristics of environment are appropriate for particular form of utilization, and methods of land evaluation are reviewed below (p. 202). An important development was achieved by Coates (1976b) when he styled engineering as the 'art or even the science of using power and materials most effectively in ways that are valuable and necessary to man' and proceeded to suggest that the geomorphologist:

> . . . must become involved in the tools of engineering because if construction causes irreparable damage to the land–water ecosystem due to lack of geomorphic input the earth scientist cannot be absolved of blame. Thus it is imperative that the geomorphic engineer be involved in the decision-making processes

that plan and manage the environment.

This led Coates (1976b, p. 6) to advocate a field of geomorphic engineering as combining

> . . . the talents of the geomorphology and engineering disciplines. It differs from environmental geology, wherein man is studied as one of the typical surface processes that change the landscape, and instead brings knowledge of physical systems to bear on systems that may require construction for their solution. The geomorphic engineer is interested in maintaining (and working towards the accomplishment of) the maximum integrity and balance of the total land–water ecosystem as it relates to landforms, surface materials and processes.

There is an outstanding need for attention to be directed towards such geomorphic engineering and it is apparent that if physical geographers do not fully respond to this call then scientists in other disciplines will!

A final group of developments concern *prediction and design*. Whereas evaluation of environment is primarily devoted to contemporary environment for particular uses, prediction and design are concerned more with the future. Although this is the most recent aspect of the development of applications of physical geography it effectively embraces several related themes. First it depends upon environmental impact statements which have been reviewed by Clark, Bisset and Wathern (1980) and by Lee (1983) and have been referred to several aspects of physical environment including the response of alluvial channels to river regulation (Hey, 1976) and of ecosystems (Moss, 1976). In environment impact assessments (EIA) the purpose is 'to determine before implementation the environmental effects of a proposed action' (Cheremisinoff and Morresi, 1977) and there has been a great increase in EIA since the National Environmental Policy Act (1969) which came into operation in the United States on 1 January 1970. Lee (1983) suggests that following the growth since 1970 there is likely to be consolidation in the 1980s in two main directions: first the extension of EIA to planning and approval of actions broader in scope than individual development projects, and secondly further improvement in quality, cost-effectiveness and practical use made of EIA studies.

A second and related aspect arises from the need to envisage alternative landscape designs. The ultimate goal of applied landscape science could be seen as the consideration of recommendations for environmental design. This has occurred in the case of river channellization (p. 207), which is the modification of river channels for the purpose of flood control, land drainage, navigation, and the reduction or prevention of erosion. Whereas Kates (1969) perceived three elements in environmental planning as contributed by the environmental science disciplines (public health, sanitary engineering, ecology and biomedical engineering); by the enviromental design disciplines (architecture, planning); and by the social and behavioural sciences (economics, political science, geography, psychology, sociology) it may be that the physical geographer has a role which is not limited to the social and behavioural sciences.

Development of these four aspects of applications of physical geography is reflected in the books which have been published since 1968 and some of these are summarized in Table 9.3. Whereas some of these books have developed as extensions of the traditional subdivisions of physical geography, others have emerged as members of particular series and yet others have been produced to refer to a specific field, and the field embracing terrain evaluation, land use planning and landscape ecology is one which has been very evident. The movement towards environmental geology which is illustrated by two books included in Table 9.3 is not completely divorced from physical geography because of the overlap between geography and geology in the USA (p. 11). Keller (1976) sees environmental geology as applied geology and more specifically as:

> . . . the application of geologic information to solving conflicts, minimizing possible adverse environmental degradation or maximizing possible advantageous conditions resulting from our use of the natural and modified environment

whereas Coates (1981) embraces similar objectives in his identification of environmental geology as that subject area which relates geology as the science of the earth to human activities. A marked absence from the range of books produced hitherto (Table 9.3) has been any major concern with landscape design and this is reflected in the structures of some of the books produced since 1970. When surveying the chapter headings of applied geomorphology for example (Table 9.4), it is evident that the dominant emphasis is upon extension of existing pure systematic research rather than development of an integrated approach which may facilitate an approach to landscape design. This reticence has not been so evident in other disciplines and, for example, McHarg (1969) published his *Design with Nature* which is now becoming more frequently referred to by physical geographers. However in landscape ecology and landscape evaluation methods there is more evidence of proposals for land use design and Verstappen (1983) in part C of his book is concerned with the problems of particular environments or domains. Even more pragmatic is the way in which Cooke, Brunsden, Doornkamp and Jones (1982) focus upon the problems posed by urban areas and urbanization in dryland areas. This book, which developed from a report prepared under the United Nations University Natural Resources Program, is relevant to more than 350 cities with a population of at least 100,000 particularly as many of these cities are expanding rapidly and are encountering environmental problems. The book shows how settling of foundation materials, weathering processes affecting building foundations and sand dunes advancing over highways, suburban developments and oasis settlements are examples of problems which arise from mismanagement or misunderstanding of geomorphological conditions. This important book shows how such problems can be

Table 9.3: Progress Related to Environmental Management

Year	Publication
1968	B.W. Atkinson *The weather business: observation, analysis, forecasting and modification*
1970	D.R. Coates (ed) *Environmental geomorphology*
	W.J. Maunder *The value of the weather*
1972	T.D. Detwyler and H.G. Marcus (ed) *Urbanisation and environment: the physical geography of the city*
	K. Smith *Water in Britain: a study in applied hydrology and resource geography*
1973	J.E. Oliver *Climate and mans environment: an introduction to applied climatology*
	W.R.D. Sewell (ed) *Modifying the weather a social assessment*
	A.G. Isachenko *Principles of landscape science and physico-geographic regionalization*
1974	R.U. Cooke and J.C. Doornkamp *Geomorphology in environmental management*
	A. Warren and F.B. Goldsmith (eds) *Conservation in practise*
	J.R. Mather *Climatology: fundamentals and applications*
	I.G. Simmons *The ecology of natural resources*
1975	K. Smith *Principles of applied climatology.*
	R.D. Hey and T.D. Davies (eds) *Science, technology and environmental management*
1976	D. R. Coates (ed) *Geomorphology and engineering*
	D.R. Coates (ed) *Urban geomorphology*
	E.D. Keller *Environmental geology*
1977	J.R. Hails (ed) *Applied geomorphology*
1978	T. Dunne and L.B. Leopold *Water in environmental planning*
1979	K.J. Gregory and D.E. Walling (eds) *Man and environmental processes*
	K. Smith and G. Tobin *Human adjustment to the flood hazard*
	R.P.C. Morgan *Soil erosion*
1980	C.C. Park *Ecology and environmental management*
	J.E. Hobbs *Applied cli matology: a study of atmospheric resources*
	D.A. Davidson *Soils and land use planning*
1981	A.S. Goudie *The human impact: mans role in environmental change*
	J.R.G. Townshend *Terrain analysis and remote sensing*
	D. Dent and A. Young *Soil survey and land evaluation*
	D.R. Coates *Environmental geology*
	J.E. Oliver *Climatology: selected applications*
1982	R.C. Craig and J.L. Craft *Applied geomorphology*
	E.M. Bridges and D.A Davidson *Principles and applications of soil geography*
	R.U. Cooke, D. Brunsden, J.C. Doornkamp and D.K.C. Jones *Urban geomorphology in drylands*
1983	H.Th. Verstappen *Applied geomorphology: geomorphology surveys for environmental development*
	A.P.A. Vink *Landscape ecology and land use*
	A. Warren and F.B. Goldsmith *Conservation in perspective*
	I. Douglas *The urban environment*
1984	L. Tufnell *Glacier hazards*

This selection of books is confined to those which have a major applied component and are largely written or edited by physical geographers although some are included by geologists in North America.

Table 9.4: Chapter structures of some applied geomorphology volumes

Cooke and Doornkamp (1974) *Geomorphology in Environmental Management*

Introduction: Geomorphology and environmental problems

1	The drainage basin in environmental management
2	Soil erosion by water
3	Soil erosion by wind
4	Rivers and river channels
5	Floodplains, fans and flooding
6	Landsliding
7	Ground-surface subsidence
8	Coasts
9	Freeze, thaw, and periglacial environments
10	Material resources
11	The destruction of natural materials by weathering
12	Landforms and techniques of scenic evaluation
13	Land systems mapping
14	Geomorphological mapping

Hails (1977) *Applied Geomorphology*, edited contributions including:
Applied geomorphology in perspective

1	Applications of weathering studies
2	The role of applied geomorphology in irrigation and groundwater studies
3	Applied geomorphology and hydrology of karst regions
4	Applied fluvial geomorphology
5	The application of soil mechanics methods to the study of slopes
6	Applied geomorphological studies in deserts – a review of examples
7	Periglacial environments
8	Terrain classification – methods, applications and principles
9	Applied geomorphology in coastal-zone planning and management

Cooke, Brunsden, Doornkamp and Jones with contributions by Griffiths, Knott, Potter and Russell (1982)
Urban Geomorphology in Drylands

1	Urban development in drylands
2	Geomorphology and planned urban development in drylands
3	Systematic mapping of geomorphology
4	Aggregate resources for the construction industry in drylands
5	Salinity, groundwater and salt weathering in drylands
6	Water and sediment problems in drylands
7	Problems of sand and dust movement in drylands

Verstappen (1983) *Applied Geomorphology: Geomorphological Surveys for Environmental Development*
A Geomorphology and the survey of environmental resources

1	Geomorphology and environmental resources
2	Geomorphology in surveying and mapping
3	The role of geomorphology in geological and soil survey
4	Geomorphology in hydrological surveys
5	Geomorphology in vegetation surveys
B	**Geomorphology and appropriate use of the natural environment**
6	Geomorphology and rural land use
7	Geomorphology and urbanization
8	Geomorphology and engineering
9	Geomorphology and mineral exploration/research
10	Geomorphology and development planning

Table 9.4—(*contd.*)

C	**Geomorphology and surveying for planned development**
11	Analytical geomorphological survey as a tool
12	Synthetic survey of terrain
13	Flood susceptibility surveys
14	Drought susceptibility and desertification surveys
15	Slope stability and erosion surveys
16	Avalanche mapping
17	Natural hazards of endogenous origin

avoided, managed or controlled; what information on ground-surface conditions is required by urban environmental managers in drylands; and how such information can be collected, presented and analysed. The authors (Cooke, Brunsden, Doornkamp and Jones, 1982):

> . . . provide a statement of value to geomorphologists, engineers and planners on the ways in which geomorphological research can assist urban development in drylands. In addressing three different audiences, the book is written from the perspective of a team of geomorphologists who do not pretend to be engineers or planners but whose experience leads them to believe that not all planners and engineers yet recognize the potential of geomorphology for saving time and money, especially if it is used in conjunction with related environmental information provided by engineering geology and soil mechanics.

This book acknowledges that its geomorphological perspective is only one way of approaching environmental data relevant to urban development in drylands but that there are also:

> . . . equally valid approaches to the study of environmental problems in urban areas of drylands adopted by other scientists, such as geologists, ecologists, pedologists and hydrologists. Indeed, different approaches can be both stimulating and compatible. What is essential in all approaches, however, is that their procedures are clear, and their products intelligible and useful to those who have the responsibility for urban development in drylands: the major problem is one of communications.

Whether such a focus which is essentially geomorphological will be vindicated in the next two decades remains to be seen. In an excellent review of applications Jones (1983) notes that the study of the interactions between human societies and natural environmental systems, although supposedly at the heart of geographical enquiry, has been neglected by British geographers and particularly by physical geographers. He ascribes this neglect to a number of reasons including general disinterest in applied work, a tendency to focus more upon other aspects of physical geography, and an inability to provide specific answers to predict future events and this was epitomized by Gould (1973, p. 271):

> Physical geography, as most of us know it, either survives as a second-rate earth science or has vanished completely. I regard it as a separate field, with problems

> so taxing that they demand the full attention of the student if anything more than a superficial acquaintance is to be achieved. Most of physical geography is totally irrelevant to human spatial organization, except at the most obvious and naive level. One does not have to wade through the last ice age to understand the Sahara is now a desiccated and inhospitable environment to man at his present stage of technology. . . .

Jones proceeds to argue that perhaps the one factor most responsible for the neglect of applied developments may have been the specialization and compartmentalization that characterized British geography for many years and this has been further exacerbated by distinct physical and human geographies. In reviewing the expansion of applied studies Jones sees the growth of applied geomorphology as the development which may have the greatest future significance for physical geography but in reviewing environments of concern, ranges broadly over the spectrum of physical geography and identifies five categories of potentially useful work that could be undertaken by physical geographers namely (1) the assessment of natural hazard impacts to influence policy; (2) environmental auditing which embraces the assessment of changes in natural environmental systems to evaluate the need to establish new practices; (3) resource assessment; (4) impact assessment concerned with predicting future changes following human–environment interactions; and (5) reviews of earlier predictions and of the success of implemented policies and projects. Many of these contributions depend upon the relations with different planning scales and management agencies and it is necessary that the geomorphologist appreciates their relation to the decision-making process as elaborated by Cooke (1982). The conclusion reached by Jones (1983, p. 454) is that:

> . . . There still exists a range of potential inter-linkages between the natural environmental and socio-economic systems which could provide an additional focus for geographical research.
> . . . the fact that geography continues to be held in relatively low esteem by many of the traditional scientific disciplines – largely because of the persistence of outmoded perceptions as to the nature and content of the subject – has naturally led physical geographers to seek greater scientific respectability and intellectual satisfaction by concentrating increasingly within their specialized areas of study and by establishing and reinforcing links with scientists in adjunct disciplines.

However such a conclusion is not echoed in another important paper published in the same year. Johnston (1983c) reviews the extent to which claims for a more integrated geography are being realized by the applied directions being undertaken in physical geography. Johnston begins from a statement on the nature of contemporary geographical scholarship in which he notes that there has been an increased emphasis upon process, that in physical geography this is best exemplified in geomorphology, hydrology and pedology, and that human geographers have been more concerned with the processes that generate mechanisms whereas physical geographers have been more restricted to the mechanisms themselves because they do not seek to contribute to physics, chemistry and biology! Johnston (1983c, p. 13) argues that failure to realize this difference may well be the reason for misunderstanding between physical and human geographers but one could of course argue that the analogue of physics, chemistry and biology in relation to physical geography is to be found in the way in which psychology, economics and psychiatry relate to human geography. Johnston (1983c) classified work by geographers in the fields of resource analysis and management into four categories namely resource appreciation, people in the physical environment, conflict over the environment which centres upon the present and future allocation of resources, and human demands on the environment which seek to understand why conflicts exist over resources and why shortages occur. He concludes that the study of resource analysis and management has in no way provided a viable modern integration of physical and human geography although bridge-building is as necessary to link physical and human geography as it is to link both parts of geography with related social or natural sciences. Johnston's fear is that too many bridges could be flimsy so that '. . . . the excursions that they carry will be little more than academic tourism', and he raises the possibility that alternatives to a weakly bound geography department in every institution could be a variety of institutional forms or could be a restructuring to emphasize the unities of the natural sciences and of the social sciences. Finally Johnston (1983c, p. 142) asks why there is a continual search for a way to counter the split between physical and human geography and ponders:

> Is it because some kind of neurosis – among physical geographers especially, who seem to be needed. . . . Is it a need to define academic territory, to try and establish some kind of overlordship with respect to other disciplines? Is it because place is so central to the study of so much of physical and human geography that there is a feeling that they must be integrated? Or is it the influence of a few dominant figures, despite claims (Stoddart, 1981) that geography is more than a discipline defined by key figures?

These two important papers (Johnston, 1983c; Jones, 1983) have been considered at length not only because they differ in their conclusions, but also because they direct attention implicitly as well as explicitly upon the focus of studies for applied geography. It is not possible to review all the ways in which applications have developed but it is desirable to outline the features of landscape ecology and mapping, evaluation and change.

Landscape ecology and mapping

Many applied approaches have been associated with the description of environment in a way which is pertinent to its utilization and management. Although there are diverse antecedents for this approach, including for example the way in which Berg (1950) recognized major physiographic zones of the USSR in a historical-genetic approach (Isachenko, 1977) which could be useful in relation to land utilization, many of the methods of environmental description have been developed outside geography and have been imported into the subject, adapted and utilized. Vink (1968; 1983) uses the term landscape ecology, although as he points out, this was first used by Troll as the result of interaction between geography (landscape) and biology (ecology), with applications to land development, regional planning and urban planning. Although it may be envisaged as an approach which interprets landscape as supporting interrelated natural and cultural systems (Vink, 1983) it is possible to envisage a major task of landscape ecology being to describe and characterize landscape according to relationships between the biosphere and the anthroposphere.

Depiction of spatial patterns of landscape ecology may be envisaged at three related levels, systematic, quantitative and integrated. At the *systematic* level an inventory can be made followed by representation of the spatial patterns created by aspects of geomorphology, climatology, soil geography, hydrology and biogeography. In each branch of physical geography, the earlier emphasis upon static, often morphological, features has been succeeded by later attempts to focus upon processes and there have often been related developments by national mapping agencies undertaken independently of physical geography. Although national agencies produce topographic maps, soil maps, geological maps and maps of superficial deposits and also provide climatic and hydrological data, it is often necessary to develop the information available in a way that is pertinent to environmental use and management. Thus the topographic map conventionally employs contours but slope maps are more fundamentally significant in relation, for example, to land use practises, and slope categories can be directly related to angles at which agricultural implements can operate (e.g. Curtis, Doornkamp and Gregory, 1965), which relate to the incidence of mass movements (e.g. Cooke, 1977), or which are appropriate for different types of building construction or for other specified activities (e.g. Cooke, Brunsden, Doornkamp and Jones, 1982, Table III.3). Approaches to new methods of depicting systematic aspects of environment have emerged, and physical geographers have considered various ways of portraying the spatial variations of environmental parameters such as those relating to water on hydrological maps as shown in Gregory and Walling (1973), and as reviewed more generally by UNESCO (1977). Perhaps most extensively developed by contributions from physical geographers have been geomorphological maps. Whereas initially morphological maps were proposed (e.g. Savigear, 1965, Gregory and Brown, 1966) these were succeeded by geomorphological maps and Bakker (1963) proposed that such maps should ideally satisfy five major requirements which include the principles of morphological characterizing, of geomorphological-genetic interpretation, of dating, of characterizing the substrate and of sedimentology or sedimentology/pedology.

It is not feasible to satisfy all five demands and the extent to which several of the five criteria of morphology, evolution, dating, lithology and sedimentology are emphasized, varies from one mapping approach to another. Geomorphological maps have been produced and published in several countries (Demek, 1972; Demek and Embleton, 1978) and in one of the most ambitious schemes in Poland physical geographers contributed to a national scheme which involved publishing maps at 1:50,000. It has also been necessary to evolve mapping schemes which are capable of international application and an approach summarized by Cooke and Doornkamp (1974) has been adapted to apply to conditions in many different areas (Cooke, Brunsden, Doornkamp and Jones, 1982). In many specific applied projects geomorphological mapping has been the central technique and in relation to problems in drylands it is envisaged that a geomorphological mapping programme would involve 13 well defined stages proceeding through familiarization with urban planning proposals, selection of the mapping town, desk study to collect background information, map and aerial photograph acquisition, air-photo interpretation, planning mapping procedure, reconnaissance fieldwork, field mapping, extrapolation by air-photo interpretation, laboratory analyses, cartography to produce the final map, report compilation, and presentation of results (Cooke, Brunsden, Doornkamp and Jones, 1982). Although the approach associated with geomorphological mapping will undoubtedly feature prominently in applications of geomorphology, the map itself may be superseded by a data base and the advent of remote sensing with improved enhanced coverage, resolution and frequency can greatly enhance the potential available (e.g. Townshend and Hancock, 1981).

Quantitative approaches now derive from the analysis of remotely sensed digital data but quantitative techniques have been available for much longer periods of time. Thus many attributes of environment have been expressed quantitatively, as in the case of many drainage basin characteristics (e.g. Gregory and Walling, 1973) and at least two other developments have been quantitative in character including the parametric approach and automated cartography. The parametric approach is one in which measurements of environmental parameters are used for the division and classification of land on the basis of selected attribute value; this is reviewed by Ollier (1977) and under the heading of general geomorphometry by Evans (1981). Although the actual parameters used vary from one study to another, King (1970) uses process, altitude, relief, dominant geology, drainage pattern, stream frequency, characteristic plan-profile, geomorphic position, dominant facet, characteristic facet, characteristic variant, and land zone, and rather more pragmatically Speight (1969) used topographic contour maps to furnish four parameters (slope angle, rate of change of slope, contour curvature, and unit catchment area) to define land elements as a basis for producing a land element map. A further quantitative approach is founded in automated cartography where a great variety of patterns may be derived from an initial survey as exemplified in the case of soil survey data (Rudeforth, 1982).

Many of these approaches embrace some degree of integration of several aspects of the physical landscape but perhaps the most *integrative* general approach is the land systems method. This approach occupies a prominent position in textbooks (e.g. Mitchell, 1973; Cooke and Doornkamp, 1974; Hails, 1977; Dent and Young, 1981; Verstappen, 1983) and has been extensively reported. The Australian Commonwealth Scientific Industrial Research Organization (CSIRO) began extensive resource surveys in 1946 in undeveloped parts of Australia and Papua, New Guinea. In subsequent years these surveys acquired information on the geology, climate, geomorphology, soils, vegetation and land use of areas and, by collating the information, designated land systems as areas or groups of areas with recurring patterns of topography, soils and vegetation and having a relatively uniform climate. A land system may be subdivided into smaller units called land units or land facets which are in turn composed of individual slopes or land elements.

The land systems approach, as utilized in Australia and Papua New Guinea, has been extensively quoted by physical geographers and related parallel methods have also been developed in other parts of the world. The Directorate of Overseas Surveys (DOS) has utilized a land systems approach particularly in Africa (e.g. Directorate of Overseas Surveys, 1968) and Isachenko (1973a) has detailed an approach used in the USSR in which the urochischa is the basic association of facies. The fundamental physical-geographical unit has a uniform bedrock, hydrological conditions, microclimate, soil, and single type of mesorelief. In eastern Europe and the USSR the association of physical characteristics is expressed in the geosystem which has been defined as the sphere of interaction between animate and inanimate nature (Sochava, Krauklis and Snytko, 1975) and in the notion of the geocomplex. Whereas many studies have focused on the homogeneity of such landscape entities it has been suggested (Mil'kov, 1979) that it is also possible to focus on paradynamic landscape systems which concentrate on mass and energy exchange rather than on internal structure.

The land system approach has been modified for application to the problems of urban and suburban areas (e.g. Grant, Finlayson, Spate and Ferguson, 1979) and a range of related approaches have been described by Vink (1983). Approaches to the quantitative delimitation of natural regions in relation to land evaluation have been reviewed by Gardiner (1976) who demonstrated the way in which grid squares can provide the basis for the data collection matrix and showed how parameterization of the drainage network can be a foundation for a numerical approach. Although the land systems approach has the advantage that it combines many aspects of environmental character and is not use-specific, it has been criticized because it does not readily relate to a particular use; because it is not easily related to results obtained from field survey (Wright, 1972); and because it reflects an emphasis upon a static view, according to the evolution of environment, rather than upon the dynamics of the environmental system. In particular Moss has advocated an alternative approach based upon how the environment works and what it does rather than merely what it is. The approach advocated (Moss, 1969a) is biocenological rather than morphogenetric and developmental, and in the case of West Africa this led to the proposal of land use vegetation systems (Moss, 1968) because it is necessary to develop a dynamic approach to environmental systems in association with the characteristics of agriculture and land use which interact with the plant–soil system (Moss, 1969b).

A range of applications has been found for approaches based upon, or related to, the land systems approach and these include relevance to military intelligence, engineering, soil survey, agriculture, forestry and land use and to regional planning (Ollier, 1977) but for some purposes more specific techniques are required and these are reviewed under the heading of land evaluation.

Environmental evaluation

Land evaluation is the estimation of the potential of land for particular kinds of use, and Dent and Young (1981) include productive uses such as arable farming, livestock production and forestry together with uses that provide services or other benefits such as water catchment areas, recreation, tourism and wildlife conservation. Therefore the essence of land evalution is the comparison of the requirements of land use with the resource potentially offered by the environment. Whereas such evaluation has the advantage that it relates to a particular form of environment use, it has the disadvantage that because the approach is use-specific it cannot therefore be utilized for other purposes and so is very labour intensive.

Progress towards land evaluation can be realized from more general land system approaches and Isachenko (1973b) distinguished three scales of landscape research for planning purposes and these were small (*c.*1: 2½ million), medium (usually 1:200,000) and large using detailed plans (e.g. 1:2,000). Because Isachenko viewed landscape mapping in stages from the inventory of geographical complexes (with identification, mapping and description), to the evaluation of complexes with respect to a particular purpose, to the prediction of changes over a given period, and thence to drafting of recommendations for use, he envisaged four types of maps. These were inventory maps, evaluation maps, prediction maps, and recommendation maps. Whereas the evaluation maps classify terrain for a particular purpose, the prediction maps indicate the modifications likely to arise and the recommendation maps show the measures which could be used to change the environment, and this was illustrated by evaluation of land with reference to tourism (Isachenko, 1973a).

Extensions of the results of integrated surveys are an admirable way of providing a land evaluation and a further approach may be achieved by the extension of systematic data sources. National surveys of rock types, of superficial deposits, and of soils have provided the bases for landscape evaluation techniques and in the case of soil survey it is possible to proceed from maps of soil bodies, to soil quality maps and soil limitation maps, and subsequently to land classification in terms of soil crop response, of present use, of use capabilities, and of recommended use (Vink, 1968; 1983). Soil survey has contributed a major component in the derivation of systems of land capability and these are well illustrated for agricultural uses by Bridges and Davidson (1982b) who show how soil maps can be used not only in relation to crop yield and to specific types of crops but also for evaluating land for agriculture at local, national and international scales. This acknowledges the use of a framework offered by the world Food and Agriculture Organization (FAO, 1976) which facilitates appreciation of land in relation to land use systems. This framework was utilized in a study of 1935 km^2 in central Malawi by Young and Goldsmith (1977). They used a conventional soil survey and aerial photographs to identify seven soil landscapes which were each evaluated in terms of six major kinds of land use so that the requirements of land use could be compared with the land qualities of the mapping units and this was the basis for a quantitative economic analysis. This gave an exemplary instance of the way in which land evaluation can lead towards an economic assessment but Young and Goldsmith (1977, p. 430) caution that:

> . . . the current trend towards translating land evaluation into economic terms is in danger of being taken too far. The validity of economic evaluation can be very short-lived. It is dependent not only on changes in costs and prices, but also on assumption about discount rates and sometimes shadow pricing, which are (to say the least) somewhat arbitrary.

Landscape evaluation can be developed from land systems or landscape ecology surveys and in the USSR this has fostered the development of landscape geochemistry particularly in relation to the extension of soil geography to accommodate the impact of technology (Glazovskaya, 1977). This has been particularly appropriate in the case of forecasts of environmental impact. The vulerability of the natural geosystems as a result of the impact of the proposed Kansk-Achinsk lignite and electric power project was analysed (Snytko, Semenov and Davydova, 1981) by identifying landscape facies with particular geochemical conditions; by detecting types of landscape geochemical barriers; by establishing the capacity of geosystems for accumulating certain elements; and by mapping topogeochores which are regions differing in spatial relations and structure of their facies groups and which can be ranked according to their vulnerability to pollutants emitted from the projected power stations.

Many other approaches to landscape evaluation have been developed from specific branches of physical geography and geomorphological mapping has been a particularly productive basis for extensions to land evaluation. Derivative maps are especially suited to regional and town planning scales and have been utilized in relation to geotechnical investigations at a new airport site and to the evaluation of areas suitable for urban expansion (Cooke, Brunsden, Doornkamp and Jones, 1982). In the use of mapping approaches it is frequently necessary to combine the fruits of several disciplines as illustrated by the Bahrain surface materials resources survey (Brunsden, Doornkamp and Jones, 1979; 1980). A survey was undertaken between 1974 and 1976 at the request of the Ministry of Development and Engineering Services, Government of Bahrain and involved a team comprising 10 geologists, 7 geomorphologists, 2 pedologists, 2 surveyors and a cartographer. The survey produced a series of maps at a scale of 1:10,000 and an extensive report so that the final volume is 'probably the most intensive and comprehensive view of the surface materials of any state within the arid lands of the world' (Brunsden, Doornkamp and Jones, 1980). Many benefits have accrued from this survey and from others undertaken by members of the same team and this has included knowledge of the consequences of environmental processes that would not otherwise have been fully appreciated. In drylands this is well illustrated by the salinity of groundwater and by salt weathering which has been shown (Cooke, Brunsden, Doornkamp and Jones, 1982) to arise from a complex hazard which depends upon the relations between local environmental conditions, the types of salt present, the nature of susceptible materials, and the design and nature of the structures built in hazardous areas.

Some methods of land evaluation have derived from approaches other than those which are synthetic or derivative and have instead been conceived as purpose-specific. This is particularly the case in the assessment of scenery and although it can be approached by synthetic methods (e.g. Linton, 1968) it has also been attacked using the method of uniqueness by Leopold (1969). Originally inspired when the Federal Power Commission studied applications for a permit to construct one or more additional dams for hydroelectric power in the Hell's Canyon area of the Snake River, Idaho, it was necessary to consider how the landscape could be ranked so that some, possibly the most unique, could be preserved from development.

The method has also been utilized in the evaluation of riverscape (Leopold and Marchand, 1968) and a related matrix method can be extended towards environmental impact (Leopold, Clarke, Hanshaw and Balsley, 1971). Methods of assessment of the aesthetic quality of landscape have included concern with perception methods and comparisons of existing methods have been undertaken (Penning-Rowsell and Hardy, 1973; Penning-Rowsell, 1981b). A range of approaches to landscape evaluation has now been taken as indicated in Table 9.5.

Table 9.5: Examples of approaches to physical landscape evaluation (developed from Gardiner and Gregory 1977).

Purpose of land evaluation	Synthetic approach		Approach based upon specific criteria
Rural land use	Land systems Urochische	Capability units	Choromorphographic maps
Scenery	Basin morphometry Grid based morphometry Point measures	Scenic quality	Uniqueness Preference techniques
Military	Integrated survey	Trafficability Strategic units	
Urban/suburban	Engineering survey Land systems	Terrain components Building suitability	

Evaluation of environment is also a clear objective of applied climatology where the evaluation of climatic regions (Barry and Perry, 1973) in relation to agriculture, water power and resources and transport (Smith, 1975; Hobbs, 1980) are obvious lines for applied development. In addition it has been possible to evaluate the way in which climate relates to human comfort in biometeorology, to building climatology which embraces concern with climates within buildings and the ability of structures to withstand hazards, and the way in which climate relates to building and leisure. It is from a range of atmospheric hazards that much has been learnt about the interaction between environment and human activity and it has been important to demonstrate the susceptibility of specific areas to risk. Although in the field of atmospheric studies the current paradigm is suggested by Thornes (1981) to be undoubtedly positivistic, he suggests that there is a new awareness of man–atmosphere concern and this can proceed from descriptive research towards atmospheric problems examined with a sound level of socioeconomic understanding. This is attempted by Ausubel and Biswas (1980) following the Task Force meeting on the nature of climate–society research in Austria in 1980, which Thornes (1981) argues, goes further than using socioeconomic terminology to describe atmosphere events by providing a platform on which to build theory so that 'it is still early days for atmospheric management.'

Impact, prediction and design

Perhaps the heart of environmental management is concerned with the future, and this may be visualized at three levels of increasing definition which may be thought of as what will happen in terms of impact; how much will happen and when – which will be the essence of prediction; and how could the environment be shaped – which is the design stage. In some cases it is impossible to separate impact from prediction and in the case of the atmosphere the most long-established concern has been with forecasting, where long-range forecasting (Barry and Perry, 1973) can supplement the short-term forecasting based upon synoptic methods. Evaluation of the economic benefits of forecasting is a useful product of research as reviewed by Maunder (1970) and Hobbs (1980). However forecasting, although long-established, is only one of the applications from climatology and Thornes (1982 p. 561) has said:

> Whenever I use the phrase 'atmospheric management' I am conscious that certain geographers raise a sceptical eye, and question as to how the atmosphere can be 'managed'. They are fully familiar with environmental management in the form of the management of woodlands or a river channel, but they fail to realize that, for instance, our daily struggle to achieve thermal comfort is an obvious example of atmospheric management.

Thornes proceeds to show how space heating allows the atmospheric manager to have a role in an examination of the current efficiency of a heating system, and in the forecasting of effective temperatures. He shows how a small investment of road surface sensors could lead to greatly improved road danger warnings and so contribute to management of the effects of atmospheric conditions on winter maintenance of roads. Forecasting of the impact of world climate changes has already been referred to (Butzer, 1980). In addition, the need continues to evaluate the risk that may be projected: Bach (1982) has shown that the most sophisticated energy scenarios available indicate that by the year 2030 the lowest projection available may lead to the warming not experienced for at least 1000 years and the highest could give a warming not experienced on earth for 6000 years.

In other branches of physical geography it is necessary to establish the consequences of future change and much can arise from an improved understanding of temporal change anticipated in chapter 8. However in some cases there are developments proposed which could lead to dramatic changes which cannot easily be visualized. Thus major interbasin transfer of water in the USSR and the planting of shelterbelts (Rostankowski, 1982) requires that consideration should continue to be give to the environmental (L'vovich, Gangardt, Sarukhanov and Berenzer, 1982) and climatic (Chubukov, Rauner, Kubshinova, Potapova and Shvareva, 1982) consequences. Although in many cases a considerable degree of certainty surrounds the range of impacts that will occur it is still difficult to predict with certainty when and how much change will take place. This was why Burkham (1981) argued that uncertainty still featured prominently in the prediction of future river channel changes. In an excellent study the effects of dams on flow regulation and hence on the channels downstream (Williams and Wolman, 1984) were analysed for the purpose of flood control, land drainage, navigation and the reduction or prevention of erosion. River channellization has been undertaken in many parts of the world and has often been implemented using engineering methods of resectioning or realigning the river channel but experience has shown that downstream of such channellization works there can be serious feedback consequences of erosion and aesthetic degradation and damage to structures and property. Therefore particularly in the USA there was been very considerable attention given to the problems induced by channellization works and four volumes were produced in 1971 (Committee on Government Operations, 1971) followed by a fifth report with additional views (Committee on Government Operations 1973) in which it was stated that:

> A common thread running through the Sub-Committtee, hearings, correspondence, and subsequent studies was not that channellization per se was evil, but rather that inadequate consideration was being given to the adverse environmental effects of channellization. Indeed there is considerable evidence that little was known about these effects and even more disturbing, little was done to ascertain them.

Experience of the adverse effects of channellization, some of which have produced 'ecological disasters', has stimulated the search for alternative methods which do not have such dramatic downstream effects and which minimize the degradation of the environmental quality and particularly the aesthetic appearance of the channellization scheme. This has been achieved by proposals for alternatives to stream channellization and these have been reviewed by Brookes, Gregory and Dawson (1983) to include stream restoration which involves minimal straightening of channels, retention of trees to promote bank stability, minimization of channel reshaping and general emulation of the morphology of natural stream channels; and stream renovation which is similar to restoration but also included water-based methods of channel maintenance. Studies have been undertaken to evaluate the success of such techniques (Keller, 1975; Nunnally, 1978) of working with the river rather than against it and in addition there have been proposals for a sympathetic contrasted with an unsympathetic, approach to riverscape design by the Nature Conservancy Council (Newbold, Purseglove and Holmes, 1983). Such approaches to environmental design have not been the prerogative of the physical geographer but one of the implications of the geomorphic engineering approach is that not only is it necessary to become more familiar with the methods used by practitioners of other disciplines but it is also desirable to assess the efficiency of alternative design strategies and it is imperative that this should proceed towards problems of environmental design. Such movements make refreshing trends and some other examples are shown in Table 9.6.

Table 9.6: Examples of design recommendations arising from research investigations

Problem	Recommendations	Source
Stream restoration	Minimal straightening of river channels and emulation of the morphology of natural stream channels	Keller and Hoffman 1976; Nunnally, 1978.
Design of road cut slopes in tropics	Models for prediction of pore-water pressures from knowledge of storm precipitation, material permeability and topography to assist in design of cut slopes.	Anderson, 1982
Forecasting off-road trafficability	Deterministic soil moisture submodel coupled to a principal empirical soil-moisture trafficability model.	Anderson, 1983
Proposal for reduction of avalanche hazard	Analysis of avalanche hazard at Ophir in mountain Colorado using historical sources and suggested measures to reduce avalanche hazard.	Ives, 1976
Highway engineering design	Use of geomorphological mapping techniques to produce recommendations about exact locations and construction of road.	Brunsden and Jones, 1975. Doornkamp, Brunsden, Jones, Cooke and Bush, 1979
Flood alleviation and land drainage	Manual of assessment techniques to enable cost benefit to be evaluated as basis for design of flood alleviation and land drainage schemes.	Penning-Rowsell and Chatterton, 1977

The future

Many of the examples in this chapter, and they are admittedly a small sample of those available, indicate how applied considerations have often developed beyond physical geography but have been adopted and developed by physical geographers. Land classification, land evaluation, and environmental impact statements have all developed initially in other disciplines. However physical geography has not only taken note of such developments but it has also extended the systematic branches by utilizing and refining prediction and modelling stategies and it has encompassed new priorities for attention. In this way there has been landscape geochemistry, environmental geomorphology and geomorphic engineering. To some extent this progress has been achieved by an enhanced view taken of consultancy and contract reseach. Penning-Rowsell (1981a, p. 11) has pointedly argued that:

> Contract research is still seen in many quarters as inferior to Research Council sponsorship which in turn is seen as inferior to unfinanced scholarship. . . . The reputation of the subject as a whole needs careful nurturing following the

> quantification debacle. An essential part of this nurturing involves a considered analysis of the potentials and problems of closer involvement by academic geographers in an advisory capacity with environmental groups, policy analysis and decision-making in the outside world.

In the progress towards greater involvement of this kind it is inevitable that specific conclusions must be reached and decisions must be taken and in introducing the first issue of *Applied Geography* the editor wrote (Briggs, 1981, p. 6) that:

> . . . the applied geographer needs to be brave. He needs to commit himself before he knows all the answers. He needs to be prepared to make public mistakes. But he must be prepared to learn from them.

In summarizing four contributions that an environmental geomorphologist has to offer to land managers Coates (1982) includes an eclectic approch to land–water ecosystems, a knowledge of feedback systems, the recognition of potential thresholds and the site-specific application of classic geomorphic principles. Coates (1982, p. 166) concludes that:

> The geomorphologist is probably the last of the science generalists, because by necessity he has had to have a proper background in not only geography and geology but also in mathematics, other sciences, and aspects of engineeering in order to understand the complex relationships that operate in the dynamics of the earth's surface. Thus, the environmental geomorphologist is in a position to not only bridge the gap with peer natural scientists but also to translate various pieces of a puzzle into a composite whole.

This view should not be interpreted to signify that environmental geomorphologists or applied physical geographers are 'super-scientists-synthesizers and integraters of everything and anything – when clearly we are not. . . .' (Johnston, 1983c, p. 143) but rather that the greater opportunities available for applications of physical geography are at the interface with, or lie between, other disciplines. Such applications have now increasingly developed as extensions from the branches of physical geography rather than from a separate field of applied physical geography. In the future such applications developing naturally as extensions of physical geography research will be more frequent, especially when catalysed by developments in remote sensing and information technology outlined in chapter 10.

Conclusion

10

Evolving eighties

Applications and other approaches in physical geography could be further advanced and enhanced by the two developments indicated at the end of the last chapter. It could be argued that in recent years physical geography has been hampered by two major constraints. First in obtaining data on environment so that it has had to resort to field survey and field monitoring which is expensive and time demanding. Dependence upon national surveys and upon national data collection programmes has increased but in many cases such data have not been collected in an appropriate way and with sufficient frequency, so that very often resurvey was an expensive necessity for the progress of physical geography research. Secondly has been the requirement for increased funding for physical geography and this has been necessary for field survey, for the purchase and operation of continuous field monitoring equipment and for the purchase and use of equipment for laboratory analysis. It is long accepted that the physical sciences including geology, and also the biological sciences demand and require high levels of laboratory and technical support but it has not been easy for physical geography to convince others that it also needs laboratories, technical support in the form of personnel of field and laboratory analysis, and hardware and subsequently software, compatible with that required in many parts of geology or biology for example. It is unfortunate that the culmination of growth of physical geography to be more technique-and process-oriented has occurred at a time when the amount of money available has ceased to increase as rapidly as in the past, with the consequence that the approach of a steadystate environment has made it more difficult to achieve an improvement in the support available for physical geography. However the two developments referred to concerned with remote sensing and information technology may have great implications for physical geography which may begin to mitigate at least to some extent these problems of data acquisition and need for increased funding.

Remote sensing

Terrestrial remote sensing encompasses all those techniques which may be used to obtain information about the earth's surface and its atmosphere by sensors which sense and record radiation from the electromagnetic spectrum and are located on specific platforms. Many such platforms are available and include aircraft, a balloon or even a large tower. Information obtained by remote sensing from the platforms of an aircraft obtaining black and white, and subsequently colour, air photographs has long been employed in physical geography and was a particularly important ingredient for terrain resources evaluation and was therefore employed in relation to landscape ecology, land systems and more specific land evaluation techniques. However two particularly important developments have related first to the electromagnetic spectrum, as more wavebands have been used, and secondly to the number of platforms that have been used. The electromagnetic spectrum includes the photographic bands which are the visible and near infrared and which have been used for more than 30 years, but it also includes linescan (visible and infrared), active microwave (side-looking radar), and passive microwave (Hardy, 1981) and there has been investigation of the way in which sensing of all parts of the electromagnetic spectrum can be used to enhance research in physical geography. A further way in which remote sensing has expanded dramatically is by increase in the number of platforms available and in particular the advent of satellite platforms greatly enhanced the potential available from remote sensing. This enhancement has been achieved because not only can satellite platforms survey large areas of the world but they can also provide frequent and repeated surveys of the same area. Thus potentially the problems of data collection and of repeated observation are alleviated by remote sensing from satellites. The multispectral scanner on the Landsat satellite system has been available since 1972 and following a sun-synchronous orbit the satellites repeat a pass along given track every eighteenth day. This means that data are available every 18 days at a particular location and that information is available on the progress of deliberate or incidental environmental change. Thus in the 1970s it was possible to deduce from Landsat imagery the progress of dam construction along the Huang He in China (Smil, 1979). Whereas air photograph interpretation has been able to make a great contribution to land and resource inventory, satellite imagery can provide an even more significant contribution in this field (Allan, 1978; Townshend, 1981a). In the field of satellite climatology significant developments have been made (Barrett, 1974) and Barrett (1970) showed how rainfall intensities could be predicted from satellite data and subsequently there have been many additional developments including the estimation of rainfall (Barrett and Martin, 1981) and the climatology of clouds (Henderson-Sellers, 1980). The fields to which satellite data analysis can be applied are now legion and Lulla (1983) has collected together the applications for Landsat data, and resolved them into those concerned with aquatic environment and ecosystems, including chlorophyll and particulate and suspended solids monitoring; with wetland and coastal ecosystems including wetland biomass estimation and measurement of coastal primary productivity; with terrestrial ecosystems including leaf area index estimations, applications to agricultural in disease prediction and to forest to estimate timber volume, and to terrestrial biomass and primary productivity estimates; and with scarred landscapes including desertification and studies of drought impact. There are obviously many implications of remote sensing data sources for physical geography and these include the possibility of taking a global view, of obtaining frequently repeated imagery, of obtaining images that allow interpretation of patterns such as plant disease that are not feasible by other methods, and by analysis of results that is essentially a non-destructive way of obtaining estimates of bioenvironmental parameters such as biomass, leaf area index, and canopy coverage (Lulla, 1983)

It is impossible to cover the potential range of applications by examples in this chapter but two may indicate the scope available. In the study of desert areas the information obtained from satellite imagery has allowed much more detailed mapping and interpretation of dune patterns and of sand seas and it has facilitated the inference of palaeoclimates (Fryberger and Goudie, 1981). Great strides have also been possible in hydrology because water has unique spectral characteristics in the visible, near infrared thermal infrared and microwave bands and also because the high specific heat of water means that it is frequently readily distinguishable by thermal energy sensors, so that most satellite sensors are of value to the hydrologist (Walling, 1983). In hydrology, as in other field, there have been extensive reviews of the potential available and achieved for remote sensing applications (e.g. Deutsch, Wiesnet and Rango, 1981). Examples of such hydrological applications arise in relation to floods because flood plain mapping, the monitoring of flood progression, and the prediction of floods via snow pack or storm observation can all be made (Ferguson, Deutsch and Kruus, 1980). It has now been shown that space-borne imaging radar (SIR) can detect drainage channels even where covered by a layer of wind-blown sand although such features would not be visible on a Landsat image (Elachi, 1983). Satellite systems may continue to be developed faster than organizations can exploit them for ongoing, useful and cost-effective applications for the benefit of the ordinary citizen (Barrett, 1981) but there are two related areas which need attention in physical geography research.

First is the need to appreciate the significance of the greater resolution becoming available and, whereas resolution in the earlier Landsat systems was of the order of 60m, the more recent developments including Landsat 4 provided a resolution of 30m. In March 1984 Landsat 5 satellite was launched and carried a high-resolution imaging system known as the 'thematic mapper'; this is a sensor which records information at six discrete wavelengths for successive areas 30 metres square on the ground together with the emitted radiation in the thermal infrared waveband in every 120 metres square and this information is recorded every 16 days for nearly all the earth's surface. In 1985 the first of a series of French earth observation satellites (SPOT-1) will be launched and will provide 10m resolution panchromatic and 20m resolution multispectral data (Briggs and Jackson, 1984). This improvement in resolution is potentially very significant because it means that the essential detail of many aspects of environmental systems can now be detected so that, for example drainage networks can be determined in sufficient detail to be of use in runoff modelling.

Second is the need for further research on the significance of the higher-resolution satellite data. All serious and detailed interpretation of satellite data must be undertaken with reference to ground control and only by the careful comparison of ground measurements and the satellite data can effective interpretation be well founded. Thus Townshend (1981b) has argued that there are three areas in which research is needed, namely that resolution measures must be more closely linked to the quality and quantity of information which can be extracted from the data; that information on spatial properties of most terrain attributes needs to be significantly improved; and that results from these two research endeavours must be integrated to allow assessment of the benefits from improvements in resolution.

Applications of remote sensing were established in the 1970s (e.g. Barrett and Curtis, 1976) and the implications of greater resolution and greater potential are being appreciated but it is now necessary to define and refine the methods that will allow this vast information source to be used to its optimum level in physical geography.

Information technology

Although the interpretation of aerial photographs was for many years done visually and one expected to see a visual photographic image, the digital processing of remotely sensed data does not necessarily involve visual images. Information obtained by remote sensing is collected for picture elements (pixcels) and stored as digital data for this framework. A single scene from the thematic mapper covering an area of 185 × 185 km comprises about 300 million separate items of information and because many such scenes may be required in analysis for one purpose the enormity of the information processing task is evident. Furthermore, in his assessment of the future developments in the handling of remote sensing data Townshend (1981c) concludes that the most important is likely to be the integration of remotely sensed information with other data sources to provide geographical information systems.

Image processing systems for the analysis of satellite digital data have been developed as microprocessors and microcomputers have advanced in capability and have decreased in size. This is the culmination of a series of developments which have proceeded from the use of computers for analysis and simulation of environmental problems, to computer cartography and thence to data bases. Simply expressed, the data base is computer-stored information which replaces the published map or data set. Whereas the latter are time-dependent and readily become outdated, the data base can be continuously updated and therefore can be the basis for the most up-to-date map or list issued when required. This requires investigation of the way in which environmental information can be stored in data base form.

The way in which stream networks can be stored for this purpose (Gardiner, 1982) and can then provide the basis for a cartographic data base (Klein, 1982) are examples of implications which have been explored.

A number of implications arise from the developments in information technology. With readily available visual display units (VDU) and associated processing systems it is possible to process information which may or may not be derived from remote sensing and thence to provide a great variety of outputs which are very versatile. In this way processing and analysis is greatly accelerated and it is possible to envisage stages and speed of analysis hitherto unprecedented (Monmonier, 1982). In cases where there is a requirement to use large volumes of spatial data of different kinds it may be necessary to develop an automated geographic information system (GIS). In reviewing computer graphics in relation to environmental planning, Teicholz and Berry (1983) have collected several edited contributions which demonstrate the potential of such geographic information systems which are simply information systems stored by referencing each item to a set of geographic coordinates. Many geographic information systems were devised in response to the needs of government programmes and the Coastal Zone Management Act of 1972 led to the development of an automated geographical information system for the State of Carolina and this has been the basis for a review of data needs for the State (Cowen, Vang and Waddell, 1983). In specific cases data bases have now been developed, and to relate to environmental assessment at a regional scale it is necessary to characterize the regional environment, to determine potential stress areas, and to evaluate possible impacts. To accomplish these objectives Olson, Klopatek and Emerson (1983) created a geoecology data base for the conterminous USA, embracing 1000 variables on file for county – subcounty units, and this provides data on terrain, water resources, forestry vegetation, wildlife, agriculture, land use, climate, air quality, population and energy. Such developments not only provide a way of achieving landscape ecology (p. 198) or providing a basis for landscape evaluation (p. 202) but they can also foster an emphasis upon the integral character of the physical environment. This may be one way of avoiding the problem that Dury (1970) perceived, that:

> . . . geographers face the choice between extinction on the one hand, and, on the other, integrated work of the kind which our predecessors claimed to be particularly suited to geography and to geographers.

Information technology has not merely assisted in the analysis, storage and retrieval of information but it can also facilitate obtaining information. This has particularly been the case where monitoring systems can be developed, capable of more continuous monitoring than ever before envisaged and also economical to produce. For example data loggers can now be assembled for comparatively low cost and this makes data more readily accessible more cheaply. It is now quite usual for environmental monitoring to produce data in the form of computer-compatible tapes so that the days of time-consuming chart analysis are disappearing. The possible impact of fifth-generation computers which will proceed towards decision making is something to be considered. Artificial intelligence may have a strong impact on geographic theory and practice and should allow the solution of problems that were previously difficult or impossible to solve (Smith, 1984). As data collection and data-sifting become even more readily facilitated it is important to remember (Rhind, 1984) that we should: 'think less of computers as tools for replicating what could be done manually and more as opportunities which permit us to tackle quite different research problems.'

Trends so far

Trends engendered by remote sensing and information technology could influence physical geography and its future, possibly more fundamentally than any development in the last century. In addition there are a number of other trends which have emerged in previous chapters which could be recalled here. It is inevitable that physical geography should continue to have a more secure scientific basis so that we should no longer expect a treatment by a physical geographer to be non-mathematical nor should we be unaware of developments in other sciences. Therefore as Moss (1979) has argued there are no

> . . . *a priori* reasons which may be advanced to contradict the contention that geography may be validly viewed as a science, and that it has potential for development as a scientific study, whether we chose to focus on the discipline as a general broad field of study embracing a number of disparate parts, or to be concerned with the study of a central theme such as regionalism.

The branches of enquiry in physical geography will have to be pursued to their logical conclusion. In the past research has often stopped too short and this has been remedied by attention devoted to processes, to human impact, to systems, to time and to applications but there is still further scope to be realized. Indeed as Briggs (1983) has argued in his assessment of applied geography:

> As yet there appear to be relatively few advances in attempts at systems modelling in the applied field. . . . This is nowhere more apparent than in the field of applied geomorphology where problem-based studies still seem to be on a relatively intuitive level and where the developments made in understanding geomorphic systems still have to be implemented in relation to real-world problems.

Perhaps physical geographers should cease riding bandwaggons and use them rather more! Thus the geomorphic bandwaggon parade (Jennings, 1973) can be envisaged to have corollaries across physical geography and there has been a temptation to absorb the jargon and to pay lip service to a particular bandwaggon without using it sufficiently. In the progress of research it is inevitable that there will be further interaction with other disciplines and this should engender greater respect for physical geography without divorcing the branches of physical geography too much one from another. At least one discipline has to focus upon the interaction of the components of the physical environment and it would seem that physical geographers should continue to direct themselves to this task. We should not be too wedded to the traditional divisions of physical geography because this tends to avoid the focus upon the interaction between the divisions and to make interest in other areas such as hydrology somewhat uncomfortable to accommodate, because they may not readily fit into the traditional four-fold division of geomorphology, climatology, soil geography and biogeography.

Greater concern for pure and applied problems of the physical environment would require a more unified physical geography but should not foster a return to eclectic ideas. Perhaps the next generation of physical geographers will finally accept geography and physical geography as disciplines analogous to any other and not as disciplines which retrieve material from others. The movement towards a more integrated physical geography may be expressed in the recent production of physical geography texts (e.g. King, 1980a; Dury, 1981; White, Mottershead and Harrison, 1984; Thompson, Mannion, Mitchell, Parry and Townshend, 1985) to succeed the texts which were devoted to one of the traditional divisions and to complement the texts produced in North America which have avoided fragmentation rather more than those in Britain. Perhaps however physical geographers have spent too much time writing books rather than research papers!

One dilemma not resolved concerns the relation of physical and human geography, and in their review of geography in the United Kingdom 1980–84 Munton and Goudie (1984) see three major themes evident in physical geography namely an historical approach, a concern for the future of the planet, and man–environment relationships related to human problems, but they also perceive that some geographers maintain that physical and human geography are becoming increasingly divergent. Although the whole of geography may require more theory (Anuchin, 1973) the disparity in the theory requirements for human and physical geography may not help to reduce the centrifugal tendencies.

Many interesting research papers have emerged from physical geography in recent years and the subject now has the scope, the foundation and the ideas so that the time is ripe for further developments. T.H. Huxley (1825–95) played an important role in physical geography at the end of the nineteenth century and his integral physical geography or physiography may be emulated again at the end of the twentieth century when physical geography has progressed through, and incorporated achievements from, a series of approaches. If this volume has offered one perspective on these approaches it will have been successful and it may help to remember the oft quoted words of T.H. Huxley:

> Science is nothing but trained and organized common sense.
>
> T.H. Huxley *Collected essays* iv The method of Zadig (London: Macmillan, 1894)

Bibliography

ACKERMANN, W.C. 1966: *Guidelines for Research on Hydrology of Small Watersheds. US Department of Interior* OWRR **26**.

ADAMS, G.F. (ed.) 1975: *Planation Surfaces. Peneplains, Pediplains and Etchplains.* Stroudsburg, Pa.: Dowden, Hutchinson & Ross, 476 pp.

AHLMANN, H.W. 1948: Glaciological research on the north Atlantic coasts. *Royal Geographical Society Research Series* No. **1,** 83 pp.

AHNERT, F. 1962: Some reflections on the place and nature of physical geography. *Professional Geographer* **14,** 1–7.

ALEXANDER, D. 1979: Catastrophic misconception? *Area* **11** (3). 228–30.

ALLAN, J.A. 1978: Remote sensing in physical geography. *Progress in Physical Geography* **2,** 55–79.

ALLEN, J.R.L. 1970: *Physical Processes of Sedimentation.* London: Allen & Unwin.

ALLEN, T.F.H. and STARR, T.B. 1982: *Hierarchy: Perspectives for Ecological Complexity,* Chicago: University of Chicago Press.

AMOROCHO, J. and HART, W.E. 1964: A critique of current methods in hydrologic systems investigation. *Transactions American Geophysical Union* **45,** 307–21.

ANDERSON, M.G. 1975: Some statistical approaches towards physical hydrology in large catchments. In PEEL, R.F., CHISHOLM, M.D.I. and HAGGETT, P. (eds.), *Processes in Physical and Human Geography, Bristol Essays* (London: Heinemann Educational), 91–109.

— 1982: Predicting porewater pressures in road cut slopes in the West Indies. *Applied Geography* 2, 55–68.

— 1983: Forecasting off-road trafficability. *Applied Geography* **3,** 239–53.

ANDREWS, J.T. 1970a: A geomorphological study of Post-Glacial uplift with particular reference to Arctic Canada. *Institute of British Geographers Special Publication* No. **2**.

— 1970b: Techniques of till fabric analysis. *British Geomorphological Research Group Technical Bulletin* **6,** 43 pp.

— 1972: Glacier power, mass balances, velocities and erosion potential. *Zeitschrift für Geomorphologie* **13,** 1–17.

— 1975: *Glacier systems: an approach to glaciers and their environments.* North Scituate, Mass.: Duxbury Press, 191 pp.

ANDREWS, J.T. and MILLER, G.H. 1980: Dating Quaternary deposits more than 10,000 years old. In CULLINGFORD, R.A., DAVIDSON, D.A. and LEWIN, J., *Timescales in Geomorphology* (Chichester: Wiley), 263–87.

ANUCHIN, V.A. 1973: Theory of geography. In CHORLEY, R.J. (ed.), *Directions in Geography,* (London: Methuen), 43–64.

APPLEBY, P.G. and OLDFIELD, F. 1978: The calculation of lead-210 dates assuming a constant rate of supply of unsupported 210 pb to the sediment. *Catena* **5,** 1–8.

ATKINSON, B.W. 1968: *The weather business: observation, analysis forecasting and modification*. London: Aldus Books.

— 1978: The atmosphere: recent observational and conceptual advances. *Geography* **63**, 283–300.

— 1979: Precipitation. In GREGORY, K.J. and WALLING, D.E. (eds.), *Man and Environmental Processes* (London: Dawson), 23–37

— 1980: Climate. In BROWN, E.H. (ed.), *Geography Yesterday and Tomorrow* (Oxford: Oxford University Press), 114–29.

— 1981: *Mesoscale Atmospheric Circulations*. London: Academic Press.

— 1983: Numerical modelling of thermally-driven mesoscale airflows involving the planetary boundary layer. *Progress in Physical Geography* **7**, 177–209.

AUSUBEL, J. and BISWAS, A.K. 1980: *Climatic Constraints and Human Activities* Oxford: Pergamon Press.

BACH, W. 1982: Future world energy development and climate risk. *Progress in Physical Geography* **6**, 549–60.

— 1984: Carbon dioxide and climatic change: an update. *Progress in Physical Geography* **8**, 83–93.

BAGLEY, D. 1966: *Wyatt's Hurricane*. London: Collins.

BAGNOLD, R.A. 1940: Beach formation by waves; some model experiments in a wave tank. *Journal Institution of Civil Engineers* **15**, 27–52.

— 1941: *The Physics of Blown Sand and Desert Dunes*. London: Methuen. 2nd edn 1954.

— 1960: Sediment discharge and stream power: A preliminary announcement. *US Geological Survey Circular* **421**.

— 1979: Sediment transport by wind and water. *Nordic Hydrology* **10**, 309–22.

BAKER, V.R. 1978a: Palaeohydraulics and hydrodynamics of scabland floods. In BAKER V.R. and NUMMEDAL, D. (eds.), *The Channeled Scabland* (Washington: NASA), 59–79.

— 1978b: Large-scale erosional and depositional features of the channeled scabland. In BAKER, V.R. and NUMMEDAL, D. (eds.), *The Channeled Scabland* (Washington: NASA), 81–115.

— 1978c: The Spokane flood controversy and the Martian outflow channels. *Science* **202**, 1249–56.

— 1981: *Catastrophic Flooding: The Origin of the Channeled Scabland*. Stroudsburg, Pa.: Dowden, Hutchinson & Ross 366 pp.

BAKKER, J.P. 1963: Different types of geomorphological maps. In *Problems of geomorphological mapping, Geographical Studies* No. **46**, Warszawa, 13–31.

BAKKER, J.P. and LE HEUX, W.N. 1946: Projective – geometric treatment of O. Lehmann's theory of the transformation of steep mountain slopes. *Proceedings Koninklijke Nederlandsche Akademie Van Wetenschappen* **49**, 533–47.

— 1952: A remarkable new geomorphological law. *Proceedings Koninklijk Nederlandsch. Akad. Wet., Series B.* **55**, 399–410, 554–71.

BALCHIN, W.G.V. 1952: The erosion surfaces of Exmoor and adjacent areas. *Geographical Journal*, **118**, 453–76.

BARBER, K.E. 1976: History of vegetation. In CHAPMAN, S., (ed.), *Methods in Plant Ecology* (Oxford: Blackwell), 5–83.

BARNES, C.P. 1954: The geographic study of soils. In JAMES, P.E. and JONES, C.F. (eds.), *American Geography Inventory and Prospect* (Syracuse University Press: Association of American Geographers), 382–95.

BARRETT, E.C. 1970: The estimation of monthly rainfall from satellite data. *Monthly Weather Review* **98**, 198–205.

— 1974: *Climatology from Satellites*. London: Methuen

—1981: Satellite rainfall estimation by cloud indexing methods for desert locust survey and control. In DEUTSCH, M., WIESNET, D.R. and RANGO, A. (eds.), *Satellite hydrology* (Minneapolis: American Water Resources Association), 92–100.

BARRETT, E.C. and CURTIS, L.F. 1976: *Introduction to Remote Sensing*. London: Chapman & Hall.

BARRETT, E.C. and MARTIN, D.W. 1981: *The Use of Satellite Data in Rainfall Monitoring*. London: Academic Press.

BARROWS, H.H. 1923: Geography as human ecology. *Annals Association of American Geographers* **13**, 1–4.

BARRY, R.G. 1963: Appendix. An introduction to numerical and mechanical techniques. In MONKHOUSE, F.J. and WILKINSON, H.R., *Maps and diagrams* (London: Methuen), 385–423.

— 1967: Models in meteorology and climatology in CHORLEY, R.J. and HAGGETT, P., *Models in Geography* (London: Methuen), 97–144.

— 1979: Recent advances in climate theory based on simple climate models. *Progress in Physical Geography* **3**, 119–31.

BARRY, R.G. and PERRY, A.H. 1973: *Synoptic Climatology. Methods and Applications*. London: Methuen.

BARRY, R.G. and CHORLEY, R.J. 1976: *Atmosphere, Weather and Climate*. London: Methuen, 3rd edn, 432 pp.

BAULIG, H. 1935: The changing sea level. *Transactions Institute of British Geographers* **3**, 1–46.

BEARDMORE, N. 1851, 1862: *Manual of Hydrology*. London: Waterlow & Sons.

BEISHON, J. and PETERS, G. 1972: *Systems Behaviour*. London: Open University/ Harper & Row.

BELASCO, H.E. 1952: *Characteristics of air masses over the British Isles*. London: HMSO.

BENNETT, H.H. 1938: *Soils and Men*. Washington DC: US Department of Agriculture Yearbook.

BENNETT, R.J. 1979: Statistical problems in forecasting long-term climate change. In WRIGLEY, N., (ed.), *Statistical Applications in the Spatial Sciences* (London: Pion), 242–56.

BENNETT, R.J. and CHORLEY, R.J. 1978: *Environmental Systems: Philosophy Analysis and Control*. London: Methuen.

BERG, L.S. 1950: *Natural Regions of the USSR*. Translated by Olga Adler, edited by MORRISON, J. and NIKIFOROFF, C.C. New York: American Council of Learned Societies.

BERGLUND, B.E. 1983: Palaeohydrological studies in lakes and mires – a palaeoecological research strategy. In GREGORY, K.J., (ed.) *Background to Palaeohydrology* (Chichester: Wiley), 237–56.

BERTALANFFY, L. VON 1972: *General Systems Theory*, New York: Braziller.

BIBBY, C. *T.H. Huxley: Scientist, Humanist and Educator*. London: Watts.

BIRD, E.C.F. 1979: Coastal processes. In GREGORY, K.J. and WALLING, D.E. (ed.), *Man and Environmental Processes* (London: Dawson), 82–101.

BIRD, J.H. 1963: The noosphere: a concept possibly useful to geographers. *Scottish*

Geographical Magazine **79**, 54–6.

BIRKELAND, P.W. 1974: *Pedology, Weathering and Geomorphological Research*. New York: Oxford University Press, 285 pp.

BIRKS, H.J.B. and BIRKS, H.H. 1980: *Quaternary Palaeoecology*. London: Arnold.

BISHOP, P. 1980: Popper's principle of falsifiability and the irrefutability of the Davisian cycle. *Professional Geographer* **32**, 310–15.

BISWAS, A. 1970: *History of Hydrology*. Amsterdam: North Holland Publishing Company.

BLONG, R.J.1982: *The Time of Darkness*. Seattle and London: University of Washington Press, 257 pp.

BOULDING, K.E. 1968: General system theory – the skeleton of science. In BUCKLEY W. (ed.), *Modern Systems Research for the Behavioural Scientist* (Chicago: Aldine).

BOULTON, G.S., JONES, A.S., CLAYTON, K.M. and KENNING, M.J. 1977: A British ice sheet model and patterns of glacial erosion and deposition in Britain. In SHOTTON, F.W. (ed.), *British Quaternary Studies: Recent Advances* (Oxford: Clarendon), 231–46.

BOURNE, R. 1931: Regional survey and its relation to stocktaking of the agricultural and forest resources of the British Empire. *Oxford Forestry Memoir* **13**.

BOWEN, D.Q. 1978: *Quaternary Geology*. Oxford: Pergamon, 221 pp.

— 1979: Geographical perspective on the Quaternary. *Progress in Physical Geography* **3**, 167–86.

BOWMAN, I. 1922: *Forest Physiography*. New York: John Wiley.

BRADLEY, W.C. 1958: Submarine abrasion and wave-cut platforms. *Bulletin Geological Society of America* 69, 967–74.

BRAITHWAITE, R.B. 1953: *Scientific Explanation*. Cambridge: Cambridge University Press.

— 1960: *Scientific Explanation*. New York: Harper Torch books.

BRETZ, J.H. 1923: The channelled scabland of the Columbia plateau. *Journal of Geology* **3**, 617–49.

BRIDGES, E.M. 1978a: Soil, the vital skin of the earth. *Geography* **63**, 354–61.

— 1978b: Interaction of soil and mankind in Britain. *Journal of Soil Science* **29**, 125–39.

— 1981: Soil geography: a subject transformed. *Progress in Physical Geography* **5**, 398–407.

BRIDGES, E.M. and DAVIDSON, D.A. (eds.)1982a: *Principles and Applications of Soil Geography*. London: Longman.

— 1982b: Agricultural uses of soil survey data. In BRIDGES, E.M. and DAVIDSON, D.A. (eds.) *Principles and applications of soil geography* (London: Longman), 171–215.

BRIGGS, D.J. 1981: Editorial. The principles and practise of applied geography. *Applied Geography* **1**, 1–8.

— 1983: Editorial. *Applied Geography* **3**, 3–4.

BRIGGS, S.A. and JACKSON, M.J. 1984: Remotely observed terrains. *The Times Higher Educational Supplement*, No. **607**, iv.

BRODA, E. 1975: *The Evolution of the Bioenergetic Processes*. Oxford: Pergamon Press, 211 pp.

BROOKES, A., GREGORY, K.J. and DAWSON, F.H. 1983: An assessment of river channelization in England and Wales. *The Science of the Total Environment* **27**, 97–111.

BROWN, E.H. 1952: The river Ystwyth, Cardiganshire: a geomorphological analysis. *Proceedings of the Geologists Association* **63**, 244–69.
— 1960: *The Relief and Drainage of Wales*. Cardiff: University of Wales Press.
— 1961: Britain and Appalachia: a study of the correlation and dating of planation surfaces. *Transactions Institute of British Geographers* **29**, 91–100.
— 1970: Man shapes the earth. *Geographical Journal* **136**, 74–85.
— 1975: The content and relationships of physical geography. *Geographical Journal* **141**, 35–48.
— 1979: The shape of Britain. *Transactions Institute of British Geographers* **NS4**, 449–62.
— 1980: Historical geomorphology – principles and practise. *Zeitschrift für Geomorphologie* Supplementband **36**, 9–15.
BROWN, E.H. and WATERS, R.S. 1974: Geomorphology in the United Kingdom since the first world war. In *Progress in Geomorphology* edited by E.H. BROWN and R.S. WATERS, London: Institute of British Geographers Special Publication No. **7**, 1–9.
BROWN, R.J.E. 1970: *Permafrost in Canada: its Influence on Northern Development*. Toronto: University of Toronto Press.
BRUNSDEN, D. 1972: Review of numerical analysis in geomorphology: An introduction. Geography **52**, 260.
BRUNSDEN, D., DOORNKAMP, J.C. and JONES, D.K.C. 1979: The Bahrain surface materials resources survey and its application to regional planning. *Geographical Journal* **145**, 1–35.
— (eds.) 1980: *Geology, geomorphology and pedology of Bahrain*. Norwich: Geo Books.
BRUNSDEN, D. and JONES, D.K.C. 1975: Large-scale geomorphological mapping and highway engineering design. *Quarterly Journal of Engineering Geology* **8**, 227–53.
BRUNSDEN, D. and THORNES, J.B. 1979. Landscape sensitivity and change. *Transactions Institute of British Geographers* **NS4**, 463–84.
BRYAN, K. 1946: Cryopedology: The study of frozen ground and intensive frost action with suggestions on nomenclature. *American Journal of Science* **244**, 622–42.
BRYAN, R.B. 1979: Soil erosion and conservation. In GREGORY, K.J. and WALLING, D.E. (eds.), *Man and Environmental Processes* (London: Butterworths), 207–24.
BUDEL, J. 1957: Die 'Doppelten Einebnungsflächen' in den feuchten Tropen. *Zeitschrift für Geomorphologie* **1**, 201–28.
— 1963: Klima-genetische Geomorphologie. *Geographische Rundschau* **15**, 269–85.
— 1969: Das system der Klima-genetischen Geomorphologie. *Erdkunde* **23**, 165–82.
— 1977: *Klima-Geomorphologie*. Berlin/Stuttgart: Borntraeger, 304 pp.
— 1980: Climatic and climatomorphic geomorphology. *Zeitschrift für Geomorphologie* Supplementband **36**, 1–8.
BUDYKO, M.I. 1958: *The Heat Balance of the Earth's Surface*. Translated by N. STEPANOVA from original dated 1956. Washington: US Weather Bureau.
BUDYKO, M.I. and GERASIMOV, I.P. 1961: The heat and water balance of the earth's surface, the general theory of physical geography and the problem of the transformation of nature. *Soviet Geography* **2**, 3–11.
BULL, P.A. 1981: Environmental reconstruction by electron microscopy. *Progress in*

Physical Geography **5**, 368–97.

BUNGE, W. 1973: The Geography. *Professional Geographer* **25**, 331–7.

BURKE, C.J. and ELIOT, F.E. 1954: The geographic study of the oceans. In JAMES, P.E. and JONES, C.F. (eds.) *American Geography Inventory and Prospect* (Syracuse University Press: Association of American Geographers), 410–27.

BURKHAM, D.E. 1981: Uncertainties resulting from changes in river form. *Proceedings American Society of Civil Engineers, Journal Hydraulics Division* **107**, 593–610.

BURROUGH, P.A. 1981: Fractal dimensions of landscapes and other environmental data. *Nature* **294**, 240–2.

— 1983: Multiscale sources of variation in soil. I. The applications of fractal concepts to nested levels of soil variation. *Journal of Soil Science* **34**, 599–620.

BURTON, I. KATES, R.W. and WHITE, G.F. 1978: *The Environment as Hazard*. New York: Oxford University Press, 240 pp.

BUTLER, B.E. 1959: *Periodic Phenomena in Landscapes as a Basis for Soil Studies*. Soil Publication **14**, CSIRO, Australia.

BUTZER K.W. 1964: *Environment and Archaeology*. London: Methuen, 524 pp.

— 1973: Pluralism in geomorphology. *Proceedings Association American Geographers* **5**, 39–43.

— 1974: Accelerated soil erosion: a problem of man–land relationships, In MANNERS, I.R. and MIKESELL, M.W., *Perspectives on Environment* (Washington, DC: Association of American Geographers), 57–77.

— Geological and ecological perspectives on the Middle Pleistocene. In BUTZER, K.W. and ISAAC, G.L.I. (eds.) *After the Australopithecines* (The Hague: Mouton), 857–74.

— 1976: *Geomorphology from the Earth*. New York: Harper & Row.

— 1980: Holocene alluvial sequences: Problems of dating and correlation. In CULLINGFORD, R.A., DAVIDSON, D.A. and LEWIN, J., *Timescales in Geomorphology* (Chichester: Wiley), 131–42.

— 1980: Adaption to global environmental change. *Professional geographer* **32**, 269–78.

— 1982: *Archaeology as Human Ecology*. Cambridge: Cambridge University Press.

CAILLEUX, A. 1947: L'indice d'emousse: definition et première application. *Comptes Rendus Sommaires de la Société Geologique de France*, 165–7.

CAINE, N. 1976: A uniform measure of subaerial erosion. *Bulletin Geological Society of America* **87**, 137–40.

CARSON. M.A. 1971: *The Mechanics of Erosion*. London: Pion.

CARSON, M.A. and KIRKBY, M.J. 1972: *Hillslope Form and Process*. Cambridge: Cambridge University Press, 475 pp.

CAWS, P. 1965: *The Philosophy of Science*. Princeton: Van Nostrand.

CHAGNON, S.A., HUFF, F.A., SCHICKEDANZ, P.T. and VOGEL, J.L. 1977: Summary of METROMEX Vol. 1: Weather Anomalies and impacts. *Bulletin 62*, State of Illinois Department Registration and Education, Illinois State Water Survey, Urbana.

CHANDLER, T.J. 1965: *The Climate of London*. London: Hutchinson.

— 1970: *The Management of Climatic Resources*. Inaugural lecture University College London.

— 1976: The Royal Commission on Environmental Pollution and the control of air pollution in Great Britain. *Area* **8**, 87–92.

CHANDLER, T.J., COOKE, R.U. and DOUGLAS, I. 1976: Physical problems of the urban environment. *Geographical Journal* **142**, 57–80.

CHAPMAN, G.P. 1977: *Human and Environmental Systems: a Geographer's Appraisal.* London: Methuen.

CHAPPELL, J.M.A. 1974: Geology of coral terraces, Huon peninsula, New Guinea: a study of Quaternary tectonic movements and sea level changes. *Bulletin Geological Society of America* **85**, 553–70.

CHARLESWORTH, J.K. 1929: The South Wales end moraine. *Quarterly Journal of the Geological Society, London* **85**, 335–58.

— 1957: *The Quaternary Era*, 2 volumes. London: Arnold.

CHEREMISINOFF, P.E. and MORRESI, A.C. 1977: *Environmental Assessment and Impact Statement Handbook*. Ann Arbor, Mich.: Ann Arbor Science.

CHISHOLM, M. 1967: General systems theory and geography. *Transactions Institute of British Geographers* **42**, 42–52.

CHORLEY, R.J. 1962: Geomorphology and general systems theory. *US Geological Survey Professional Paper* **500–B**, 1–10.

— 1965: A re-evaluation of the geomorphic system of W.M. Davis. In CHORLEY, R.J. and HAGGETT, P. (eds.), *Frontiers in Geographical Teaching* (London: Methuen), 21–38.

— 1966: The application of statistical methods to geomorphology. In DURY, G.H. (ed.), *Essays in Geomorphology* (London: Heinemann). 275–387.

— 1967: Models in geomorphology. In CHORLEY, R.J. and HAGGETT, P.(eds.), *Models in Geography* (London: Methuen), 59–96.

— (ed.) 1969a: *Water, Earth and Man*, London: Methuen.

— 1969b: The drainage basin as the fundamental geomorphic unit. In CHORLEY, R.J. (ed.) *Water, Earth and Man* (London: Methuen), 77–100.

— 1971: The role and relations of physical geography. *Progress in Geography* **3**, 87–109.

— (ed.) 1972: *Spatial Analysis in Geomorphology*. London: Methuen, 393 pp.

— (ed.) 1973: *Directions in Geography*. London: Methuen, 331 pp.

— 1973: Geography as human ecology. In CHORLEY, R.J. (ed.), *Directions in Geography* (London: Methuen), 155–69.

— 1978: Bases for theory in geomorphology. In EMBLETON, C., BRUNSDEN, D. and JONES, D.K.C. (eds.), *Geomorphology. Present Problems and Future Prospects* (Oxford: Oxford University Press), 281 pp.

CHORLEY, R.J., DUNN, A.J. and BECKINSALE, R.P. 1964: *The History of the Study of Landforms, Vol. I, Geomorphology before Davis*. London: Methuen, 678 pp.

CHORLEY, R.J., BECKINSALE, R.P. and DUNN, A.J. 1973: *The History of the Study of Landforms, Vol. II, The life and work of William Morris Davis*. London: Methuen, 874 pp.

CHORLEY, R.J. and HAGGETT, P. (eds.) 1965: *Frontiers in Geographical Teaching*. London: Methuen, 379 pp.

— (eds.) 1967: *Models in Geography*. London: Methuen.

CHORLEY, R.J. and KATES, R.W. 1969: Introduction. In CHORLEY, R.J. (ed.), *Water, Earth and Man* (London: Methuen), 1–7.

CHORLEY, R.J. and KENNEDY, B.A. 1971: *Physical Geography: A Systems Approach*. London: Prentice Hall.

CHRISTIAN, C.S. and STEWART, G.A. 1953: Survey of the Katherine-Darwin region 1946. *CSIRO Land Research Series* **1**, Melbourne.

CHUBUKOV, L.A., RAUNER, Yu.L., KUVSHINOVA, K.V., POTAPOVA, L.S., and SHVAREVA, Yu. N. 1982: Predicting the climatic consequences of the interbasin transfer of water in the Midland region of the USSR. *Soviet Geography* **22**, 426–44.

CHURCH, M. 1980: Records of recent geomorphological events. In CULLINGFORD, R.A., DAVIDSON, D.A. and LEWIN, J., *Timescales in Geomorphology* (Chichester: John Wiley), 13–29.

— 1984: On experimental method in geomorphology. In BURT, T.P. and WALLING, D.E. (eds.), *Catchment Experiments in Fluvial Geomorphology* (Norwich: Geobooks), 563–80.

CLAPPERTON, C.M. 1972: Patterns of physical and human activity on Mount Etna. *Scottish Geographical Magazine* **88**, 160–7.

CLARK, B.D., BISSET, R. and WATHERN, P. 1980: *Environmental Impact Assessment*. London: Mansell.

CLARK. M.J. 1978: Geomorphology in coastal zone management. Geography **63**, 273–82.

CLARK, M.J. and GREGORY, K.J. 1982: Physical geography techniques: a self-paced course. *Journal of Geography in Higher Education* **6**, 123–31.

CLAYTON, K.M. 1970: The problem of field evidence in geomorphology. In OSBORNE, R.H., BARNES, F.A. and DOORNKAMP, J.C. (eds.), *Geographical Essays in Honour of K.C. Edwards* (University of Nottingham: Geography Department).

— 1971: Geomorphology – a study which spans the geology/geography interface. *Journal of the Geological Society* London, **127**, 471–6.

— 1980a: Geomorphology. In BROWN, E.H. (ed.), *Geography Yesterday and Tomorrow* (Oxford: Oxford University Press), 167–80.

— 1980b: Beach sediment budgets and coastal modification. *Progress in Physical Geography* **4**, 471–86.

— 1984: Review of 'The Urban Environment'. *The Times Higher Education Supplement*, 17 February 1984, 26.

CLEMENTS, F.E. 1916: *Plant Succession, an Analysis of the Development of Vegetation*. Washington: Carnegie Institution.

CLIMAP Project Members 1976: The surface of the ice-age earth. *Science* **191**, 1131–7.

COATES, D.R. (ed.) 1971: *Environmental Geomorphology*. Binghamton: State University of New York Publications in Geomorphology.

— (ed.) 1972: *Environmental Geomorphology and Landscape Conservation Volume 1. Prior to 1900*. Stroudsburg: Dowden, Hutchinson & Ross.

— 1973 (ed.) *Environmental Geomorphology and Landscape Conservation Volume* III *Non-urban regions* Stroudsburg: Dowden, Hutchinson & Ross, 483 pp.

— (ed.) 1976a: *Geomorphology and Engineering* Stroudsburg: Dowden, Hutchinson & Ross.

— 1976b: Geomorphic engineering. In *Geomorphology and Engineering* (Stroudsburg: Dowden, Hutchinson & Ross), 3–21.

— 1976c *Urban Geomorphology*. Geological Society of America Special Paper **174**.

— 1980: Subsurface influences. In GREGORY, K.J. and WALLING, D.E. (eds.), *Man and Environmental Processes* (London: Butterworths), 163–88.

— 1981: *Environmental Geology*. Chichester: Wiley.

— 1982: Environmental geomorphology perspectives. In FRAZIER, J.W (ed.), *Applied Geography Selected Perspectives* (Englewood Cliffs, NJ: Prentice Hall), 139–69.

COATES, D.R. and VITEK, J.D. (eds.) 1980: *Thresholds in Geomorphology*. London: Allen & Unwin.
COFFEY, W.J. 1981: *Geography. Towards a General Spatial Systems Approach*. London: Methuen, 270 pp.
COLE, J.M. and KING, C.A.M. 1968: *Quantitative Geography: Techniques and Theories in Geography*. London: John Wiley.
COLE, M.M. 1963: Vegetation and geomorphology in Northern Rhodesia: An aspect of the distribution of savanna of central Africa. *Geographical Journal* **129**, 290–310.
COLLINSON, J. and LEWIN, J. (eds.) 1983: *Modern and Ancient Fluvial Systems*. International Association of Sedimentologists special publication no. **6**, Oxford: Blackwells, 575 pp.
COMMITTEE ON GOVERNMENT OPERATIONS, SUB-COMMITTEE ON CONSERVATION AND NATURAL RESOURCES 1971: *Stream channelization*: hearing before 92nd Congress, First Session, Washington, DC, 4 vols.
COMMITTEE ON GOVERNMENT OPERATIONS 1973: *Stream channelization; what federally financed draglines and bulldozers do to our nations streams*. Fifth report together with additional views, Washington, DC.
COMMONER, B. 1972: *The Closing Circle: Confronting the Environmental Crisis*. London: Cape.
CONACHER, A.J. and DALRYMPLE, J.B. 1977: The nine-unit landsurface model: an approach to pedogeomorphic research. *Geoderma* **18**, 1–154.
COOKE, R.U. 1971: Systems and physical geography. *Area* **3**, 212–16.
— 1976: *An Empty Quarter*, An inaugural lecture. Bedford College, London 20 pp.
— 1977: Applied geomorphlogical studies in deserts: a review of examples. In HAILS, J.R. (ed.), *Applied Geomorphology* (Amsterdam: Elsevier), 183–225.
— 1982: The assessment of geomorphological problems in dryland urban areas. *Zeitschrift für Geomorphologie* Supplementband **44**, 119–28.
COOKE, R.U., BRUNSDEN, D. DOORNKAMP, J.C. and JONES, D.K.C. 1982: *Urban Geomorphology in Drylands*. Oxford University Press, 324 pp.
COOKE, R.U. and DOORNKAMP, J.C. 1974: *Geomorphology in Environmental Management*. Oxford: Oxford University Press.
COOKE, R.U. and REEVES, R.W. 1976: *Arroyos and Environmental Change in the American South-West*. Oxford: Clarendon Press.
COOKE, R.U. and WARREN, A. 1973: *Geomorphology in Deserts*. London: Batsford.
COOPE, G.R., MORGAN, A. and OSBORNE, P.J. 1971: Fossil coleoptera as indicators of climatic fluctuations during the last Glaciation in Britain. *Palaeogeography, Palaeoclimatology, Palaeoecology* **10**, 87–101.
COSTA, J.E. and GRAF, W.L. 1984: The geography of geomorphologists in the United States. *Professional Geographer* **36**, 82–9.
COTTON, C.A. 1922: *Geomorphology of New Zealand*. Wellington: Dominion Museum.
— 1942: *Landscape as Developed by the Processes of Normal Erosion*. Cambridge University Press.
— 1948: *Climatic Accidents*. Wellington: Whitcomb and Tombs.
COURT, A. 1957: Climatology: complex, dynamic and synoptic. *Annals Association of American Geographers* **47**, 125–36.
COURTNEY, F.M. and NORTCLIFF, S. 1977: Analysis techniques in the study of soil

distribution. *Progress in Physical Geography* **1**, 40–64.
COWEN, D.J., VANG, A.H. and WADDELL, J.M. 1983: Beyond hardware and software: Implementing a state-level geographical information system. In TEICHOLZ, E. and BERRY, B.J.L., *Computer Graphics and Environmental Planning* (Englewood Cliffs, NJ: Prentice Hall), 30–51.
COWLES, H.C. 1911: The causes of vegetative cycles. *Annals Association of American Geographers* **1**, 3–20.
CRAIG, R.C. and CRAFT, J.L. 1982: *Applied Geomorphology*. Binghamton: State University of New York.
CROIZAT, 1978: Deduction, induction and biogeography. *Systematic Zoology* **27**, 209–13.
CRUICKSHANK, J.G. 1972: *Soil geography*. Newton Abbot: David & Charles.
CULLINGFORD, R.A., DAVIDSON, D.A. and LEWIN, J. (eds.) 1980: *Timescales in Geomorphology*. Chichester: Wiley.
CURTIS, C.D. 1976: Chemistry of rock weathering: fundamental reactions and controls. In DERBYSHIRE E. (ed.) *Geomorphology and Climate* (London: John Wiley), 25–58.
CURTIS, L.F., DOORNKAMP, J.C. and GREGORY, K.J. 1965: The description of relief in field studies of soils. *Journal of Soil Science* **16**, 16–30.
CURTIS, L.F. and SIMMONS, I.G. 1976: Man's Impact on Past Environments. *Transactions Institute of British Geographers* **NS1**, 1–384.
CZECH, H. and BOSWELL, K.C. 1953: Translation of PENCK, W., *Morphological Analysis of Landforms*. London: Macmillan.
CZUDEK, T. and DEMEK, K.J. 1970: Thermokarst in Siberia and its influence on the development of lowland relief. *Quaternary Research* **1**, 103–20.
DALYRMPLE, J.B., CONACHER, A.J. and BLONG, R.J. 1969: A nine-unit hypothetical landsurface model. *Zeitschrift für Geomorphologie* **12**, 60–76.
DANIELS, R.B., GAMBLE, E.E. and CADY, J.G. 1971: The relationship between geomorphology and soil morphology and genesis. *Advances in Agronomy* **23**, 51–88.
DANSEREAU, P. 1957: Biogeography An Ecological Perspective. New York: Ronald Press.
DARBY. H.C. 1956: The clearing of the woodland in Europe. In THOMAS, W.L. (ed.), *Man's Role in Changing the Face of the Earth* (Chicago: University of Chicago Press). 183–216.
DAVID, T. 1958: Against geography. *Universities Quarterly* **12**, 261–73.
DAVIDSON, D.A. 1980a: Erosion in Greece during the first and second millennia BC. In CULLINGFORD, R.A., DAVIDSON, D.A. and LEWIN, J. (eds.), *Timescales in Geomorphology* (Chichester: Wiley). 143–58.
— 1980b: *Soils and Land Use Planning*. London: Longman.
DAVIDSON, D.A. and SHACKLEY, M.L. 1976: *Geoarchaeology: Earth Science and the Past*. London: Duckworth.
DAVIES, G.L. 1968: *The Earth in Decay: A History of British Geomorphology 1578–1878*. London: MacDonald Technical & Scientific, 390 pp.
DAVIES, J.L. 1969: *Landforms of Cold Climates*. Cambridge, Mass. MIT Press.
— 1973: *Geographical Variation in Coastal Development*. New York: Hafner, 204 pp.
DAVIS, J.C. 1973: *Statistics and Data Analysis in Geology*. New York: Wiley.
DAVIS, W.M. 1912: *Die erklärende Beschreibung der Landformen*, translated by A.

Ruhl. Leipzig: Teubner.

DE GEER, G. 1912: A geochronology of the last 12000 years. *Eleventh International Geological Congress* Stockholm, Compte Rendu 1, 241–58.

DEMEK, J. (ed.) 1972: *Manual of Detailed Geomorphological Mapping*. Prague: IGU Commission on Geomorphological Survey and Mapping.

—— 1973: Quaternary relief development and man. *Geoforum* **15,** 68–71.

DEMEK, J. and EMBLETON, C. (eds.) 1978: *Guide to Medium-Scale Geomorphological Survey and Mapping*. Stuttgart: IGU Commission on Geomorphological Survey and Mapping.

DENT, D. and YOUNG, A. 1981: *Soils and Land Use Planning*. London: Allen & Unwin.

DE PLOEY, J. (ed.) 1983: *Rainfall Simulation, Runoff and Soil Erosion. Catena* Supplement **4,** Braunschweig.

DERBYSHIRE, E. 1962: Late-glacial drainage in part of North-East Wales. An alternative hypothesis. *Proceedings Geologists Association* **73,** 327–34.

—— (ed.) 1973: *Climatic Geomorphology*. Geographical Readings: London: Macmillan, 296 pp.

—— (ed.) 1976: *Geomorphology and Climate*. Chichester: Wiley.

—— 1983: On the morphology, sediments and origin of the loess plateau of central China. In GARDINER, R. and SCOGING, H., *Mega-Geomorphology* (Oxford: Clarendon Press). 172–94.

DETWYLER, T.R. 1971: *Man's Impact on Environment*. New York: McGraw Hill.

DETWYLER, T.R. and MARCUS, M.G. (eds.) 1972: *Urbanization and Environment: The Physical Geography of the City*. Belmont, Cal.: Duxbury Press.

DEUTSCH, M., WIESNET, D.R. and RANGO, A. 1981: *Satellite Hydrology*. Minneapolis: American Water Resource Association.

DIRECTORATE OF OVERSEAS SURVEYS 1968: The land resources of Lesotho. *Land Resources Division, Land resources study* **3.**

DOBBIE, C.H. and WOLF, P.O. 1953: The Lynmouth flood of August 1952. *Proceedings Institution Civil Engineers* **2,** 522–88.

DOORNKAMP, J.C., BRUNSDEN, D., JONES, D.K.C., COOKE, R.U. and BUSH, P.R. 1979: Rapid geomorphological assessments for engineering. *Quarterly Journal of Engineering Geology* **12,** 189–204.

DOORNKAMP, J.C., GREGORY, K.J. and BURN, A.S. 1980: *Atlas of Drought in Britain 1975–76*. London: Institute of British Geographers.

DOORNKAMP, J.C. and KING, C.A.M. 1971: *Numerical Analysis in Geomorphology*. London: Arnold, 372 pp.

DOUGLAS, I. 1969: Man, vegetation and the sediment yields of rivers. *Nature* **215,** 925–8.

—— 1972: *The Environment Game*. Inaugural lecture University of New England: Armidale, New South Wales.

—— 1973: Water resources. In DAWSON, J.A. and DOORNKAMP, J.C. (eds.), *Evaluating the Human Environment* (London: Arnold), 57–87.

—— 1980: Climatic geomorphology. Present-day processes and landform evolution. Problems of interpretation. *Zeitschrift für Geomorphologie* Supplementband **36,** 27–47.

—— 1981: The city as an ecosystem. *Progress in Physical Geography* **5,** 315–67.

—— 1982: The unfulfilled promise: earth surface processes as key to landform evolution. *Earth Surface Processes and Landforms* **7,** 101.

—— 1983: *The Urban Environment*. London: Arnold.

DUNNE, T. and LEOPOLD, L.B. 1978: *Water in Environmental Planning*. San Francisco: W.H. Freeman, 818 pp.

DURY, G.H. 1964a: Principles of underfit streams. *US Geological Survey Professional Paper* **452A**.

— 1964b: Subsurface exploration and chronology of underfit streams. *US Geological Survey Professional Paper* **452B**.

— 1965: Theoretical implications of underfit streams. *US Geological Survey Professional Paper* **452C**.

— 1970: Merely from nervousness. *Area* **4**, 29–32.

— 1972: Some recent views on the nature, location, needs and potential of geomorphology. *Professional Geographer* **24**, 199–202.

— 1980a: Neocatastrophism: A further look. *Progress in Physical Geography* **4**, 391–413.

— 1980b: Step-functional change in precipitation at Sydney. *Australian Geographical Studies* **18**, 62–78.

— 1981: *An Introduction to Environmental Systems*. London: Heinemann, 366 pp.

— 1982: Step-functional analysis of long records of streamflow. *Catena* **9**, 379–96.

DYLIK, J. 1952: The concept of the periglacial cycle in Middle Poland. *Bulletin de la Société des Sciences et des lettres de Lodz* **3**.

EDEN, M.J. 1964: The savanna ecosystem, southern Rupununi, British Guiana. *McGill University, Savanna Research Series vol. 1.*.

— 1974: The origin and status of savanna and grassland in southern Papua. *Transactions Institute of British Geographers* **63**, 97–110.

EDWARDS, K.C. 1964: The importance of biogeography. *Geography* **49**, 85–97.

EDWARDS, K.J. 1983: Quaternary palynology: consideration of a discipline. *Progress in Physical Geography* **7**, 113–25.

ELACHI, E. 1983: Radar images of earth from space. *Science* **218**, 46–53.

ELLIS, J.B. 1979: The nature and sources of urban sediments and their relation to water quality: a case study from north-west London. In HOLLIS, G.E. (ed.), *Man's Impact on the Hydrological Cycle in the United Kingdom* (Norwich: Geobooks), 199–216.

EMBLETON, C.E. 1961: The geomorphology of the Vale of Conway, North Wales, with particular reference to its deglaciation. *Transactions Institute of British Geographers* **29**, 47–70.

— 1964: Sub-glacial drainage and supposed ice-dammed lakes in North-East Wales. *Proceedings Geologists Association* **75**, 31–8.

— 1982: *Glaciology in the Service of Man*. Inaugural lecture, Kings College London.

EMBLETON, C. and KING, C.A.M. 1968: *Glacial and Periglacial Geomorphology*. London: Arnold, 608 pp.

— 1975: *Periglacial Geomorphology*. London: Arnold, 203 pp.

EMBLETON, C. and THORNES, J. (eds.) 1979: *Process in Geomorphology*. London: Arnold, 436 pp.

EVANS, C.D.R. and HUGHES, M.J. 1984: The Neogene succession of the south western approaches, Great Britain. *Journal Geological Society London* **141**, 315–26.

EVANS, I.S. 1981: General geomorphometry. In GOUDIE, A.S. (ed.) *Geomorphological Techniques* (London: Allen & Unwin), 31–7.

EYRE, S.R. 1978: *The Real Wealth of Nations*. London: Arnold.

FAIRBRIDGE, R.W. 1961: Eustatic changes in sea level. *Physics and Chemistry of the*

Earth **5**, 99–185.
—— 1980: Thresholds and energy transfer in geomorphology. In COATES, D.R. and VITEK, J.D. (eds.) *Thresholds in Geomorphology* (London: Allen & Unwin), 43–50.
FAO 1976: *A Framework for Land Evaluation*. Rome: FAO Soils Bulletin **32**.
FELS, E. 1965: Nochmals: Anthropogen Geomorphologie, *Petermanns Geographische Mitteilungen* **109**, 9–15.
FENNEMAN, N.M. 1931: *Physiography of the Western United States*. New York: McGraw Hill.
—— 1938: *Physiography of the Eastern United States*. New York: McGraw Hill.
FERGUSON, H.L., DEUTSCH, M. and KRUUS, J. 1980: Applications to floods of remote sensing from satellites. In SALOMONSON, V.V. and BHAVSAR, P.D. (eds.) *The Contribution of Space Observations to Water Resources Management*. Advances in Space Exploration, New York: Pergamon 195–205.
FISHER, C.A. 1970: Whither regional geography. *Geography* **55**, 373–89.
FLEMING, N.C. 1964: Form and function of sedimentary particles. *Journal of Sedimentary Petrology* **35**, 381–90.
FLENLEY, J.R. 1979: *The Equatorial Rain Forest: a Geological History*. London: Butterworth.
FLINT, R.F. 1947: *Glacial and Pleistocene Geology*. New York: Wiley.
—— 1957: *Glacial and Pleistocene Geology*. New York: Wiley.
FOSBERG, F.R. (ed.) 1963: *Man's Role in the Island Ecosystem*. Honolulu: Bishop Museum Press.
FREEMAN, T.W. 1980: The Royal Geographical Society and the development of Geography. In BROWN, E.H. (ed.) *Geography Yesterday and Tomorrow* (Oxford, Oxford University Press), 1–99.
FROSTICK, L. and REID, I. 1977: The origin of horizontal laminae in ephemeral stream channel-fill. *Sedimentology* **24**, 1–9.
FRYBERGER, S. and GOUDIE, A.S. 1981: Arid geomorphology. *Progress in Physical Geography* **5**, 420–8.
FURLEY, P.A. and NEWEY, W.W. 1983: *Geography of the Biosphere, An Introduction to the Nature Distribution and Evolution of the World's Life Zones*. London: Butterworths, 413 pp.
GARDINER, V. 1976: Land evaluation and the numerical delimitation of natural regions. *Geographica Polonica* **34**, 11–30.
—— 1982: Stream networks and digital cartography. In BICKMORE, D.P. (ed.), *Perspective in the Alternative Cartography, Cartographica* No. **19**, Monograph 28, 38–44.
GARDINER, V. and GREGORY, K.J. 1977: Progress in portraying the physical landscape. *Progress in Physical Geography* **1**, 1–22.
GARDNER, R. and SCOGING, H. 1983: *Mega-geomorphology*. Oxford: Clarendon Press.
GATES, W.L. 1976: Modelling the ice-age climate. *Science* **191**, 1138–44.
GEIGER, R. 1965: *The Climate near the Ground*, Second edition. Cambridge, Mass: Harvard University Press.
GERARD, R.W. 1964: Entitation, animorgs, and other systems. In MESAROVIC, M.D. (ed.), *Views on General Systems Theory* (New York: Wiley), 119–24.
GERASIMOV, I.P. 1961: The moisture and heat factors of soil formation *Soviet Geography* **2**, 3–12.
—— 1968: Constructive geography: aims, methods and results. *Soviet Geography* **9**, 739–55.

—— 1984: The contribution of constructive geography to the problem of optimization of society's impact on the environment. *Geoforum* **15**, 95–100.

GERRARD, A.J. 1981: *Soils and Landforms*. London: George Allen & Unwin.

GERSMEHL, P.J. 1976: An alternative biogeography. *Annals Association of American Geographers* **66**, 223–41.

GILBERT, G.K. 1877: Report on the geology of the Henry Mountains. *US Geological Survey*, Rocky Mountain Region Report, 160 pp.

—— 1914: The transportation of debris by running water. *US Geological Survey Professional Paper* **86**, 263 pp.

GILES, B.D. 1978: Wyatt's Hurricane: a meteorological critique. *Weather* **33**, 384–92.

GJESSING, J. 1978: Glacial geomorphology: present problems and future prospects. In EMBLETON, C., BRUNSDEN, D. and JONES, D.K.C., *Geomorphology. Present Problems and Future Prospects* (Oxford: Oxford University Press), 107–31.

GLAZOVSKAYA, N.A. 1977: Current problems in the theory and practice of landscape geochemistry. *Soviet Geography* **18**, 363–73.

GOETZ, A.F.H. 1980: Geological remote sensing in the 1980s. In SIEGAIL, B. and GILLESPIE, A.R., *Remote Sensing in Geology* (Chichester: John Wiley & sons), 679–85.

GOLOMB, B. and EDER, H.M. 1964: Landforms made by man. *Landscape* **14**, 4–7.

GOUDIE, A.S. 1977: *Environmental Change*. Oxford: Clarendon Press, 1st edn, 258 pp, 2nd edn 1983.

—— (ed.) 1981a: *Geomorphological Techniques*. London: George Allen & Unwin.

—— 1981b: *The Human Impact, Man's Role in Environmental Change*. Oxford: Blackwell, 316 pp.

—— 1983: Dust storms in space and time. *Progress in Physical Geography* **7**, 502–30.

GOUDIE, A.S. and BULL, P. 1984: Slope process change and colluvium deposition in Swaziland: an SEM analysis. *Earth Surface Processes and Landforms* **9**, 289–99.

GOULD, P. 1973: The open geographic curriculum. In CHORLEY, R.J. (ed.), *Directions in Geography* (London: Methuen) 253–84.

GRAF, W.L. 1977: The rate law in fluvial geomorphology. *American Journal of Science* **277**, 178–91.

—— 1979a: Catastrophe theory as a model for change in fluvial systems. In RHODES, D.D. and WILLIAMS, G.P. (eds.), *Adjustments of the fluvial system* (Dubuque, Iowa: Kendall/Hunt), 13–32.

—— 1979b: Mining and channel response. *Annals Association of American Geographers* **69**, 262–75.

—— 1982: Spatial variation of fluvial processes in semi-arid lands. In THORN, C.E. (ed.), *Space and Time in Geomorphology* (London: Allen & Unwin). 193–217.

—— 1983: Downstream changes in stream power in the Henry Mountains, Utah. *Annals Association of American Geographers* **73**, 373–87.

—— 1984: The geography of American field geomorphology. *Professional Geographer* **36**, 78–82.

GRAF, W.L., TRIMBLE, S.W., TOY, T.J. and COSTA, J.E. 1980: Geographic geomorphology in the eighties. *Professional Geographer* **32**, 279–84.

GRANT, K., FINLAYSON, A.A., SPATE, A.P. and FERGUSON, T.G. 1979: Terrain analysis and classification for engineering and conservation purposes of the Port Clinton area, Queensland, including the Shoalwater Bay Military training area, *Division of Applied Geomechanics Technical Paper No.* **29**, Melbourne: CSIRO.

GREGORY, D. 1978: *Ideology, Science and Human Geography*. London: Hutchinson, 198 pp.
GREGORY, K.J. 1965: Proglacial lake Eskdale after sixty years. *Transactions Institute of British Geographers* **36**, 149–62.
— 1976: Changing drainage basins. *Geographical Journal* **142**, 237–47.
— (ed.) 1977: *River Channel Changes*. Chichester: Wiley.
— 1978a: A physical geography equation. *National Geographer* **12**, 137–41.
— 1978b Valley carved in the Yorkshire Moors. *Geographical Magazine* **50**, 276–9.
— 1978c: *Down to Earth: Four Dimensions of Physical Geography*, inaugural lecture, University of Southampton, 30 pp.
— 1978d: Fluvial processes in British basins. In EMBLETON, C., BRUNSDEN, D. and JONES. D.K.C. (eds.), *Geomorphology Present Problems and Future Prospects* (Oxford: Oxford University Press), 40–72.
— 1981: River channels. In GREGORY, K.J. and WALLING, D.E., *Man and Environmental Processes* (London: Butterworth), 123–43.
— (ed.) 1983: *Background to Palaeohydrology*. Chichester: Wiley, 486 pp.
GREGORY, K.J. and BROWN, E.H. 1966: Data processing and the study of land form. *Zeitschrift für Geomorphologie* **10**, 237–63.
GREGORY, K.J. and WALLING, D.E. 1973: *Drainage Basin Form and Process*. London: Arnold 458 pp.
— (eds.) 1979: *Man and Environmental Processes*. London: Dawson, 276 pp.
GREGORY, K.J. and WILLIAMS, R.F. 1981: Physical geography from the newspaper, *Geography* **66**, 42–52.
GREGORY, S. 1963: *Statistical Methods and the Geographer*. London: Longman.
— 1976: On geographical myths and statistical fables. *Transactions Institute of British Geographers* **NS1**, 385–400.
— 1978: The role of physical geography in the curriculum. *Geography* **63**, 251–64.
GRIGORYEV, A.Z. 1961: The heat and moisture regions and geographic zonality. *Soviet Geography* **2**, 3–16.
GROVE, A.T. 1977: Desertification. *Progress in Physical Geography* **1**, 296–310.
GUPTA, A. 1983: High magnitude floods and stream channel response. In COLLINSON, J.D. and LEWIN, J. (eds.), *Modern and Ancient Fluvial Systems* (Oxford: Blackwell), 219–27.
GVODETSKIY, N.A., GERENCHUK, K.I., ISACHENKO, A.G. and PREOBRAZHENSKIY, V.S. 1971: The present state and future tasks physical geography. *Soviet Geography* **12**, 257–66.
HACK, J.T. 1960: Interpretation of erosional topography in humid temperate regions. *American Journal of Science* **258**, 80–97.
HAGGETT, P. 1977: Geography in a steady-state environment. *Geography* **62**, 159–67.
HAGGETT, P. and CHORLEY, R.J. 1969: *Network Analysis in Geography*. London: Arnold, 348 pp.
HAINES-YOUNG, R.H. and PETCH, J.R. 1980: The challenge of critical rationalism for methodology in physical geography. *Progress in Physical Geography* **4**, 63–77.
HAILS, J.R. (ed.) 1977: *Applied Geomorphology*. Amsterdam: Elsevier.
HANNA, L.W. 1983: Agricultural meteorology. *Progress in Physical Geography* **7**, 329–44.
HARDY, J.R. 1981: Data collection by remote sensing for land resources survey. In

TOWNSHEND, J.R.G. (ed.), *Terrain Analysis and Remote Sensing* (London: George Allen & Unwin).

HARE, F.K. 1951a: Geographical aspects of meteorology. In TAYLOR, G. (ed.), *Geography in the Twentieth Century* (New York: Philosophical Library), 178–95.

— 1951b: Climatic classification. In STAMP, L.D. and WOOLDRIDGE, S.W. (eds.) *London Essays in Geography, Rodwell Jones Memorial Volume* (London: University of London), 111–34.

— 1957: The dynamic aspects of climatology. *Geografiska Annaler* **39**, 87–104.

— 1965: Energy exchanges and the general circulation. *Geography* **50**, 229–41.

— 1966: The concept of climate. *Geography* **5**, 99–110.

— 1969: Environment: Resuscitation of an idea. *Area* **4**, 52–5.

— 1973: Energy-based climatology and its frontier with ecology. In CHORLEY, R.J. (ed.), *Directions in Geography* (London, Methuen), 171–92.

— 1980: The planetary environment: fragile or sturdy. *Geographical Journal* **146**, 379–95.

— 1983: Climate on the desert fringe. In GARDNER. R. and SCOGING, H., *Mega-Geomorphology* (Oxford: Clarendon Press) 134–51.

HARRIS, D.R. 1965: *Plants, Animals and Man in the Outer Leeward Islands, West Indies. An Ecological Study of Antigua, Barbuda and Anguilla*. Berkeley and Los Angeles: University of California Press.

— 1968: Recent plant invasions in the arid and semi-arid southwest of the United States. *Annals Association of American Geographers* **56**, 408–22.

HARRISON, C.M. 1980: Ecosystem and communities: patterns and processes. In GREGORY, K.J. and WALLING, D.E. (eds.), *Man and Environmental Processes* (Folkstone: Dawson), 225–40.

HARRISON CHURCH, R.J. 1951: The French school of geography. In TAYLOR, G. (ed.), *Geography in the Twentieth Century* (New York: Philosophical Library), 70–90.

HARTSHORNE, R. 1939: *The Nature of Geography, a Critical Survey of Current Thought in the Light of the Past*. Lancaster, Pa: Association of American Geographers.

— 1959: *Perspective on the Nature of Geography*. Chicago: Rand McNally & Co.

HARVEY, D.W. 1969: *Explanation in Geography*. London: Arnold, 521 pp.

HAYS, J.D. and MOORE, T.C. 1973: CLIMAP Program. *Quaternary Research* **3**, 1–2.

HEATHCOTE, R.L. 1965: *Back of Bourke: a Study of Land Appraisal and Settlement in Semi-Arid Australia Carlton, Victoria:* Melbourne University Press.

HENDERSON-SELLERS, A. 1980: Albedo changes – surface surveillance from satellites. *Climatic Change* **2**, 275–81.

HEWITT, K. (ed.) 1983: *Interpretations of Calamity from the Viewpoint of Human Ecology*. Boston: Allen & Unwin, 304 pp.

HEWITT K. and BURTON, I. 1971: *The Hazardousness of a Place: a Regional Ecology of Damaging Events*. Toronto: University of Toronto Press, 154 pp.

HEWITT, K. and HARE, F.K. 1973: *Man and Environment: Conceptual Frameworks*. Association of American Geographers, Commission on College Geography, Resource Paper No. **20**, 39 pp.

HEY, R.D. 1976: Impact prediction in the physical environment. In O'RIORDAN, T. and HEY, R.D. (eds.), *Environmental Impact Assessment* (Farnborough: Saxon House), 71–81.

— 1978: Determinate hydraulic geometry of river channels. *Proceedings American Society of Civil Engineers, Journal Hydraulics Division* **104**, 869–85.

HEY, R.D. and DAVIES, T.D. (eds.) 1976: *Science Technology and Environmental Management*. Farnborough: Saxon House.

HIGGINS, C.G. 1965: Causes of relative sea-level changes. *American Scientist* **53**, 464–76.

— 1975: Theories of landscape development: A perspective. In MELHORN, W.N. and FLEMAL, R.C. (eds.), *Theories of Landform Development* (Binghamton: State University of New York Publications in Geomorphology), 1–28.

HILL, A.R. 1975: Biogeography as a sub-field of geography. *Area* **7**, 156–60.

HILLS, T.L. 1965: Savannas: A review of a major research problem in tropical geography. *Canadian Geographer* **9**, 216–28.

— 1974: The savanna biome; a case study of human impact on biotic communities. In MANNERS, I.R. and MIKESWELL, M.W. (eds.) *Perspectives on Environment* (Washington: Association of American Geographers), 342–74.

HJULSTRØM, F. 1935: Studies of the morphological activity of rivers as illustrated by the River Fyris. *Bulletin Geological Institute Uppsala* **25**, 221–527.

HOBBS, J.E. 1980: *Applied Climatology: a Study of Atmospheric Resources*. London: Buttterworth.

HOLLIS, G.E. 1975: The effect of urbanization on floods of different recurrence intervals. *Water Resources Research* **11**, 431–4.

— (ed.) 1979: *Man's Impact on the Hydrological Cycle in the United Kingdom*. Norwich: Geobooks.

HOLMES, C.D. 1941: Till fabric. *Bulletin Geological Society of America* **52**, 1299–354.

HOLZNER, L. and WEAVER, G.D. 1965: Geographic evaluation of climatic and climatogenetic geomorphology. *Annals Association of American Geographers* **55**, 592–602.

HOLT-JENSEN, A. 1981: *Geography: Its History and Concepts*. London: Harper & Row, 171 pp.

HOOKE, J.M. and KAIN, R.J.P. 1982: *Historical Change in the Physical Environment: a Guide to Sources and Techniques*. London: Butterworth, 236 pp.

HORTON, R.E. 1932: Drainage basin characteristics. *Transactions American Geophysical Union* **13**, 350–61.

— 1933: The role of infiltration in the hydrologic cycle. *Transactions American Geophysical Union* **14**, 446–60.

— 1945: Erosional development of streams and their drainage basins: hydrophysical approach to quantitative morphology. *Bulletin Geological Society of America* **56**, 275–370.

HUGGETT, R.J. 1975: Soil landscape systems: a model of soil genesis. *Geoderma* **13**, 1–22.

— 1976a: Lateral translocation of soil plasma through a small valley basin in the Northaw Great Wood, Hertfordshire. *Earth Surface Processes* **1**, 99–109.

— 1976b: A scheme for the science of geography, its systems, laws and models. *Area* **8**, 25–30.

— 1980: *Systems Analysis in Geography*. Oxford: Clarendon Press.

— 1982: Models and spatial patterns of soils. In BRIDGES, E.M. and DAVIDSON, D.A. (eds.), *Principles and Applications of Soil Geography* (London: Longman), 132–70.

HUTTON, J. 1795: *Theory of the Earth*. Edinburgh, 2 volumes.

INTERNATIONAL ASSOCIATION FOR HYDROLOGICAL SCIENCES (IAHS) 1974: *Effects of Man on the Interface of the Hydrological Cycle with the Physical Environment*.

International Association for Hydrological Sciences Publication **113**.

ISACHENKO, A.G. 1973a: *Principles of Landscape Science and Physico - Geographic Regionalization*. Trans. from Russian by J.S. Massey, Melbourne: University of Melbourne Press.

— 1973b: On the method of applied landscape research. *Soviet Geography* **14**, 229–43.

— 1977: L.S. Berg's landscape - geographic ideas, their origins and their present significance. *Soviet Geography* **18**, 13–18.

IVES, J.D. 1976: Natural hazards in mountain Colorado. *Annals Association of American Geographers* **66**, 129–44.

JACKS, G.V. and WHYTE, R.O. 1939: *The Rape of the Earth*. London: Faber & Faber.

JAMES, P.E. and MARTIN, C.J. 1978: *The Association of American Geographers. The First Seventy-Five Years 1904–1979*. Association of American Geographers, 279 pp.

JARRETT, R.D. and COSTA, J.E. 1983: Multidisciplinary approach to the flood hydrology of foothill streams in Colorado. *International Symposium on Hydrometeorology*, Bethesda, Maryland, American Water Resources Association, 565–9.

JENNINGS, J.N. 1966: Man as a geological agent. *Australian Journal of Science* **28**, 150–6.

— 1973: 'Any milleniums today, Lady?' The geomorphic bandwaggon parade. *Australian Geographical Studies* **11**, 115–33.

JENNY, H. 1941: *Factors of Soil Formation, A System of Quantitative Pedology*. New York: McGraw Hill.

— 1961: Derivation of state factor equations of soils and ecosystems. *Proceedings Soil Science Society of America* **25**, 385–8.

JOHNSON, D.W. 1921; *Battlefields of the World War*. New York: American Geographical Society Research Series No. 3.

— 1931: *Stream sculpture on the Atlantic Slope*. New York: Columbia University Press.

JOHNSON, W.M. 1963: The pedon and the polypedon, *Proceedings Soil Science Society of America* **27**, 212–15.

JOHNSTON, R.J. 1979: *Geography and Geographers. Anglo American Human Geography since 1945*. London: Arnold, 232 pp.

— 1983a: *Geography and Geographers. Anglo American Human Geography since 1945* London: Arnold, 2nd edn.

— 1983b: *Philosophy and Human Geography*. London: Arnold, 152 pp.

— 1983c: Resource analysis, resource management, and the integration of physical and human geography. *Progress in Physical Geography* **7**, 127–46.

JONES, D.K.C. 1980: British applied geomorphology: an appraisal. *Zeitschrift für Geomorphologie* Supplementband **36**, 48–73.

— 1983: Environments of concern. *Transactions Institute of British Geographers* **NS8**, 429–57.

JONES, O.T. 1952: Discussion of the erosion surfaces of Exmoor and adjacent areas. *Geographical Journal* **118**, 473–4.

KATES, R.W. 1969: Comprehensive environmental planning. In HUFSCHMIDT. M.M. (ed.) *Regional Planning: Challenge and Prospects* (New York: Praeger).

KELLER, E.A. 1975: Channelization: a search for a better way. *Geology* **3**, 246–8.

— 1976: *Environmental Geology*. Columbus: C.E. Merrill Publishing Company.

KELLER, E.A. and HOFFMAN, E.K. 1976: A sensible alternative to stream channelization. *Public Works*, October, 70–2.

KENDALL, P.F. 1902: A system of glacier lakes in the Cleveland Hills. *Quarterly Journal of the Geological Society, London* **58**, 471–571.

KENNEDY, B.A. 1979: A naughty world. *Transactions Institute of British Geographers* **NS4**, 550–8.

KIDSON, C. 1982: Sea level changes in the Holocene. *Quaternary Science Reviews* **1**, 121–51.

KIMBALL, D. 1948: Denudation chronology. The dynamics of river action. *Occasional Paper No.* **8**, *University of London, Institute of Archaeology*, 21 pp.

KINCER, J.B. and others 1941: *Climate and Man*. Washington, DC: Department of Agriculture Yearbook.

KING, C.A.M. 1959: *Beaches and Coasts*. London: Arnold (2nd edn 1972).

— 1962: *Introduction to Oceanography*. London: Arnold. 2nd edn 1975: *Introduction to Physical and Biological Oceanography*, 2 vols.

— 1963: Some problems concerning marine planation and the formation of erosion surfaces. *Transactions Institute of British Geographers* **33**, 29–43.

— 1966: *Techniques in Geomorphology*. London: Arnold, 342 pp.

— 1980a: *Physical Geography*. Oxford: Blackwell, 332 pp.

— 1980b: Thresholds in glacial geomorphology. In COATES. D.R. and VITEK, J.D. (eds.), *Thresholds in Geomorphology* (London: Allen & Unwin), 297–322.

KING, C.A.M. and McCULLAGH, M.J. 1971: A simulation model of a complex recurved spit. *Journal of Geology* **79**, 22–36.

KING, L.C. 1950: A study of the world's plainlands: a new approach in geomorphology. *Quarterly Journal of the Geological Society London* **106**, 101–27.

— 1953: Canons of landscape evolution. *Bulletin Geological Society of America* **64**, 721–52.

— 1962: *The Morphology of the Earth*. Edinburgh: Oliver & Boyd.

KING, P.B. and SCHUMM, S.A. 1980: *The Physical Geography (Geomorphology) of W.M. Davis*. Norwich: Geo Books, 217 pp.

KING, R.B. 1970: A parametric approach to land system classification *Geoderma* **4**, 37–46.

KIRKBY, M.J. 1980: The streamhead as a significant geomorphic threshold. In COATES, D.R. and VITEK, J.D. (eds.), *Thresholds In Geomorphology* (London: Allen & Unwin), 53–73.

KLEIN, P.M. 1982: Cartographic data bases and river networks. In BICKMORE, D.P. (ed.) *Perspectives in the Alternative Cartography, Cartographica* No. **19**, Monograph 28, 45–53.

KNOWLES, R.L. 1974: *An Ecological Approach to Urban Growth*. Cambridge, Mass.: MIT Press, 198 pp.

KNOX, J.C. 1972: Valley alluviation in southwestern Wisconsin. *Annals Association of American Geographers* **62**, 401–10.

KRCHO, J. 1978: The spatial organization of the physical-geographical sphere as a cybernetic system expressed by means of measures of entropy. *Acta Facultatis Rerum Naturalium Universitatis Comenianal, Geographica No.* **16**, Bratislava, 57–147.

KRINSLEY, D.H. and DOORNKAMP, J.C. 1973: *Atlas of Quartz Sand Surface Textures*. Cambridge: Cambridge University Press.

KROPOTKIN, P. 1983: The teaching of physiography. *Geographical Journal* **2**, 350–9.

KRUMBEIN, W.C. 1941: The effect of abrasion on the size and shape and roundness of rock fragments. *Journal of Geology* **49**, 482–520.

KRUMBEIN, W.C. and GRAYBILL, F.A. 1965: *An Introduction to Statistical Models in Geology*. New York: McGraw Hill, 475 pp.

KUCHLER, A.W. 1954: Plant Geography. In JAMES, P.E. and JONES, C.F. (eds.) *American Geography Inventory and Prospect* (Sycracuse University Press: Association of American Geographers). 428–41.

KUHN, T.S. 1962, 1970: *The Structure of Scientific Revolutions*. Chicago, University of Chicago Press.

KUKLA, G.J. 1975: Loess stratigraphy of central Europe. In BUTZER, K.W. and ISAAC, G.Ll. *After the Australopithecines* (The Hague: Mouton). 99–188.

LAKATOS, I. 1970: Falsification and the methodology of scientific research programmes. In LAKATOS, I. and MUSGRAVE, H. (eds.), *Criticism and the Growth of Knowledge* (Cambridge: Cambridge University Press).

— 1978: *The methodology of scientific research programmes*, Philosophical Papers volume 1. Edited by J. WORRALL and G. CURRIE, Cambridge: Cambridge University Press, 250 pp.

LAMB, H.H. 1950: Types and spells of weather in the British Isles. *Quarterly Journal Royal Meteorological Society* **76**, 393–429.

LANE, F.W. 1966: *The Elements Rage*. Newton Abbott: David & Charles.

LANGBEIN, W.B. *et al.* 1949: Annual runoff in the United States. *US Geological Survey Circular* **52**, 14 pp.

LAZLO, E. 1972a: *The Systems View of the World*. New York: Braziller.

— 1972b: *Introduction to Systems Philosophy*. London: Gordon & Breach.

LEE, N. 1983: Environmental impact assessment: a review. *Applied Geography* **3**, 5–27.

LEIGHLY, J. 1954: Climatology. In JAMES, P.E. and JONES, C.F. (ed.), *American Geography Inventory and Prospect* (Syracuse University Press: Association of American Geographers), 334–61.

LEOPOLD, L.B. 1969: Landscape esthetics. *Natural History*, Oct., 37–44.

— 1973: River channel change with time: an example. *Bulletin Geological Society of America* **84**, 1845–60.

LEOPOLD, L.B., CLARKE, F.E., HANSHAW, B.B. and BALSLEY, J.R. 1971: *A Procedure for Evaluating Environmental Impact*. US Geological Survey Circular **645**.

LEOPOLD, L.B. and EMMETT, W.W. 1965: Vigil network sites: a sample of data for permanent filing. *Bulletin International Association of Scientific Hydrology* **10**, 12–21.

LEOPOLD, B. and LANGBEIN, W.B. 1962: The concept of entropy in landscape evolution. *US Geological Survey Professional Paper* **500-A**, 20 pp.

— 1963: Association and indeterminacy in geomorphology. In ALBRITTON, C.C. (ed.), *The Fabric of Geology* (Reading, Mass.: Addison-Wesley,) 184–92.

LEOPOLD, L.B. and MARCHAND, M.O. 1968: On the quantitative inventory of riverscape. *Water Resources Research* **4**, 709–17.

LEOPOLD, L.B., WOLMAN, M.G. and MILLER, J.P. 1964: *Fluvial Processes in Geomorphology*. San Francisco: Freeman, 522 pp.

LEWIN, J. 1980: Available and appropriate timescales in geomorphology. In CULLINGFORD, R.A., DAVIDSON, D.A. and LEWIN, J., *Timescales in Geomorphology* (Chichester: Wiley), 3–10.

LEWIS, W.V. 1949: The function of meltwater in cirque formation: A reply. *Geographical Review* **39**, 110–28.

— (ed.) 1960: *Investigations on Norwegian Cirque Glaciers*. London: Royal Geographical Society.

LEWIS, W.V. and MILLER, M.M. 1955: Kaolin model glaciers. *Journal Glaciology* **2**, 533–8.

LIKENS, G.E., BORMANN, F.H. PIERCE, R.S., EATON, J.S. and JOHNSTON, N.M. 1977: *Biogeochemistry of a forested ecosystem*. Berlin: Springer-Verlag.

LINDEMAN, R.L. 1942: The trophic-dynamic aspect of ecology. *Ecology* **23**, 399–418.

LINSLEY, R.K., KOHLER, M.A. and PAULHUS, J.L.H. 1949: *Applied Hydrology*. New York: McGraw Hill.

LINTON, D.L. 1951: The delimitation of morphological regions. In STAMP, L.D. and WOOLDRIDGE, S.W. (eds.) *London Essays in Geography* (London: LSE), 199–218.

— 1957: The everlasting hills. *Advancement of Science* **14**, 58–67.

— 1965: The geography of energy. *Geography* **50**, 197–228.

— 1968: The assessment of scenery as a natural resource. *Scottish Geographical Magazine* **84**, 219–38.

LOCKWOOD, J.G. 1977: Long-term climatic changes. *Progress in Physical Geography* **1**, 104–13.

— 1979a: *Causes of Climate*. London: Arnold, 260 pp.

— 1979b: Causative factors in climatic fluctuations. *Progress in Physical Geography* **3**, 111–18.

— 1980: Milankovitch theory and ice ages. *Progress in Physical Geography* **4**, 79–87.

— 1983a : Modelling climatic change. In GREGORY, K.J. (ed.) *Background to Palaeohydrology* (Chichester: Wiley), 25–50.

— 1983b: The influence of vegetation on the earth's climate. *Progress in Physical Geography* **7**, 81–9

— 1984: The southern oscillation and El Nino. *Progress in Physical Geography* **8**, 102–10.

LOWENTHAL, D. 1961: Geography, experience and imagination: Towards a geographical epistemology. *Annals Association of American Geographers* **51**, pp. 241–60.

— (ed.) 1965: *Man and Nature by George Perkins Marsh*. Cambridge, Mass.: Harvard University Press.

LUKE, J.C. 1974: Special solutions for nonlinear erosion problems, *Journal Geophysical Research* **79**, 4035–40.

LULLA, K. 1983: The Landsat satellites and selected aspects of physical geography. *Progress in Physical Geography* **7**, 1–45.

L'VOVICH, M.I., GANGARDT, G.G., SARUKHANOV, G.L., and BERENZER, A.S. 1982: Territorial redistribution of streamflow within the European USSR. *Soviet Geography* **22**, 391–405.

LYELL, C.W. 1830: *Principles of Geology*. London: Murray.

MacARTHUR, R.H. and WILSON, E.O. 1967: *The Theory of Island Biogeography*. Princeton: Princeton University Press.

McHARG, I.L. 1969: *Design with Nature*. New York: Natural History Press.

MABBUTT, J.A. 1976: Report on activities of the IGU Working Group on desertification in and around arid lands. *Geoforum* **7**, 147–52.

MALTBY, E. 1975: Numbers of soil micro-organisms as ecological indicators of changes resulting from moorland and reclamation on Exmoor. *Journal of Biogeography* **2**, 117–36.

MANDELBROT, B.B. 1977: *Fractals, form, chance and dimension*. London: Freeman.

MANLEY, G. 1952: *Climate and the British Scene*. London: Collins.

MANNERFELT, C.M. 1945: Nagra glacialmorfologiska formelement. *Geografiska Annaler* **27**, 3–239.

MANNERS, I.R. and MIKESELL, M.W. 1974: *Perspectives on Environment*. Association of American Geographers Publication No. **13**.

MARBUT, C.F. 1925: Translation of K.D. Glinka, *The Great Soil Groups of the World and their Development*. Ann Arbor: Edward Bros.

MARCUS, M.G. 1979: Coming full circle: Physical Geography in the twentieth century. *Annals Association of American Geographers* **69**, 521–32.

MARGALEF, R. 1968: *Perspectives in Ecological Theory*. Chicago: University of Chicago Press, 111 pp.

MARSH, G.P. 1864: *Man and Nature or Physical Geography as Modified by Human Action*. New York: Charles Scribner.

MASTERMAN, M. 1970: The nature of a paradigm. In LAKATOS, I. and MUSGRAVE, A. (eds.) *Criticism and the Growth of Knowledge* (Cambridge: Cambridge University Press).

MATHER, J.R. 1974: *Climatology: Fundamentals and Applications*. New York: McGraw Hill.

MATHER, J.R., FIELD, R.T., KALKSTEIN, L.S. and WILLMOTT, C.J. 1980: Climatology: The challenge for the eighties. *Professional Geographer* **32**, 285–92.

MATHER, P.M. 1976: *Computational Methods of Multivariate Analysis in Physical Geography*. London: Wiley, 532 pp.

MAUNDER, W.H. 1970: *The Value of the Weather*. London: Methuen.

MAY, R.M. 1977: Thresholds and breakpoints in ecosystems with a multiplicity of stable states. *Nature* **269**, 471–7.

MEAD, D.W. 1919: *Hydrology The Fundamental Basis of Hydraulic Engineering*. New York: McGraw Hill.

MEADE, R.H. and TRIMBLE, S.W. 1974: Changes in sediment loads in rivers of the Atlantic drainage of the United States since 1900. In *Effects of Man on the Interface of the Hydrological Cycle with the Physical Environment*, International Association of Scientific Hydrology Publication No. **113**, 99–104.

MEIGS, P. III 1954: The geographic study of water on the land. In JAMES, P.E. and JONES, C.F. (eds.) *American Geography*. Inventory and Prospect (Syracuse University Press. Association of American Geographers), 396–409.

MEINZER, O.E. (ed.) 1942: *Hydrology*. New York: McGraw Hill, 712 pp.

MIL'KOV, F.N. 1979: The contrastivity principle in landscape geography. *Soviet Geography* **20**, 31–40.

MILLER, A.A. 1931: *Climatology*. London: Methuen.

MILLER, R.L. and KAHN, J.S. 1962: *Statistical Analysis in the Geological Sciences*. New York: Wiley, 483 pp.

MILNE, G. 1935: Some suggested units of classification and mapping, particularly for East African soils. *Soil Research* **4**, 183–98.

MITCHELL, C.W. 1973: *Terrain Evaluation*. London: Longman.

MONMONIER, M.S. 1982: *Computer-assisted Cartography Principles and Prospects*. Englewood Cliffs, NJ: Prentice Hall.

MOORE, J.J., FITZSIMMONS, P., LAMBE, E. and WHITE, J. 1970: A comparison and evaluation of some phytosociological techniques. *Vegetatio* **20**, 1–20.

MORE, R.J. 1967: Hydrological models and geography. In CHORLEY, R.J. and HAGGETT, P. (ed.), *Models in Geography* (London: Methuen), 145–85.

MORGAN, R.P.C. 1979: *Soil Erosion*. London: Longman.

MORISAWA, M.E. 1968: *Streams: Their Dynamics and Morphology*. New York: McGraw Hill, 175 pp.

MOSLEY, M.P. 1982: Slopes and slope processes. *Progress in Physical Geography* **6**, 115–21.

MOSLEY, M.P. and O'LOUGHLIN, C. 1980: Slopes and slope processes. *Progress in Physical Geography* **4**, 97–106.

MOSLEY, M.P. and ZIMPFER, G.L. 1978: Hardware models in geomorphology. *Progress in Physical Geography* **2**, 438–61.

MOSS, B. 1976: Ecological considerations in the preparation of environmental impact statements. In O'RIORDAN, T. and HEY, R.D. *Environmental impact assessment* (Farnborough: Saxon House), 82–90.

MOSS, R.P. 1968: Land use, vegetation and soil factors in south west Nigeria: a new approach. *Pacific Viewpoint* **9**, 107–27.

— 1969a: The appraisal of land resources in tropical Africa. *Pacific Viewpoint* **10**, 18–27.

— 1969b: The ecological background to land-use studies in tropical Africa, with special reference to the west. In THOMAS, M.F. and WHITTINGTON, G.W. (eds.), *Environment and Landuse in Africa* (London: Methuen), 193–240.

— 1970: Authority and charisma: criteria of validity in geographical method. *South African Geographical Journal* **52**, 13–37.

— 1979: On geography as science. *Geoforum* **10**, 223–33.

MUMFORD, L. 1931: *The Brown Decades: A study of the Arts in America 1865–1895*. New York: Dover.

MUNTON, R.J.C. and GOUDIE, A.S. 1984: Geography in the United Kingdom 1980–1984. *Geographical Journal* **150**, 27–47.

MURTON, B.J. 1968: Mapping the immediate pre-European vegetation on the east coast of the North Island of New Zealand. *Professional Geographer* **20**, 262–4.

NATURE and RESOURCES 1981: CO_2 and the effects of human activities on climate: a global issue. *Nature and Resources* **17 (3)**, 2–5.

NELSON, G. and PLATNICK, N.I. 1981: *Systematics and Biogeography: Cladistics and Vicariance*. New York: Columbia University Press.

NELSON, G. and ROSEN, D.E. (eds.) 1981: *Vicariance Biogeography: a Critique*. New York: Columbia University Press.

NERC 1976: *Research in Applied and World Climatology*. Natural Environment Research Council Publications Series 'B', no. **17**, 46 pp.

NEWBOLD, C., PURSEGLOVE, J. and HOLMES, J. 1983: *Nature Conservation and River Engineering*. Shrewsbury: Nature Conservancy Council.

NEWSON, M.D. 1979: Framework for field experiments in mountain areas of Great Britain. *Studia Geomorphologica Carpatho – Balcanica* **8**, 163–74.

NIKIFOROFF, C.C. 1949: Weathering and soil evolution. *Soil Science* **67**, 219–30.

NIKIFOROFF, C.C. 1959: Reappraisal of the soil: pedogenesis consists of transactions in matter and energy between the soil and its surroundings. *Science* **129**, 3343, 186–96.

NORTCLIFF, S.M. 1984: Spatial analysis of soil. *Progress in Physical Geography* **8**, 261–9.

NUNNALLY, N.R. 1978: Stream-renovation: an alternative to channelization. *Environmental Pollution Management* **2**, 403–11.

NYE, J.F. 1952: The mechanics of glacier flow. *Journal of Glaciology* **2**, 82–93.

ODUM, E.P. 1975: *Ecology*, 2nd edn. New York: Holt Reinehart & Winston.

ODUM, H.T. and ODUM, E.C. 1976: *Energy Basis for Man and Nature*. New York: Wiley.

OKE, T.R. 1978: *Boundary Layer Climates*. London: Methuen.

OLDFIELD, F. 1977: Lakes and their drainage basins as units of sediment-based ecological study. *Progress in Physical Geography* **1**, 460–504.

— 1983a: The role of magnetic studies in palaeohydrology. In GREGORY, K.J. (ed.), *Background to Palaeohydrology* (Chichester: Wiley), 141–65.

— 1983b: Man's impact on the environment: some recent perspectives. *Geography* **68**, 245–56.

OLDFIELD, F., BATTARBEE, R.W. and DEARING, J.A. 1983: New approaches to recent environmental change. *Geographical Journal* **149**, 167–81.

OLIVER, J.E. 1973: *Climate and man's Environment: an Introduction to Applied Climatology*. New York: Wiley.

— 1981: *Climatology: Selected Applications*. New York: Halsted Press.

OLLIER, C.D. 1977: Terrain classification: methods, applications and principles. In HAILS, J.R. (ed.) *Applied Geomorphology*, (Amsterdam: Elsevier), 277–316.

— 1979: Evolutionary geomorphology of Australia and Papua-New Guinea. *Transactions Institute of British Geographers* **NS4**, 516–39.

— 1981: *Tectonics and Landforms*, Edinburgh: Oliver & Boyd, 324 pp.

OLSON, R.J., KLOPATEK, J.M. and EMERSON, C.J. 1983: Regional environmental analysis and assessment utilizing the geoecology data base. In TEICHOLZ, E. and BERRY, B.J.L., *Computer Graphics and Environmental Planning* (Englewood Cliffs, NJ: Prentice Hall), 102–18.

O'SULLIVAN, P.E. 1979: The ecosystem watershed concept in the environmental sciences – a review. *International Journal of Environmental Sciences* **13**, 273–81.

PAIN, C.F. 1978: Landform inheritance in the central highlands of Papua New Guinea. In DAVIES, G.L. and WILLIAMS, M.A.J. (eds.) *Landform Evolution in Australia* (Canberra: Australian National University Press) 5–47.

PARK, C.C. 1980: *Ecology and Environmental Management*. Folkstone: Dawson.

— Man, river systems and environmental impacts. *Progress in Physical Geography* **5**, 1–13.

PARKER, D.J. and HARDING, D.M. 1979: Natural hazard evaluation, perception and adjustment. *Geography* **64**, 307–16.

PATTON, P.C. and SCHUMM, S.A. 1975: Gully erosion, northern Colorado: a threshold phenomenon. *Geology* **3**, 88–90.

PEARSON, M.G. 1978: Snowstorms in Scotland 1831 to 1861. *Weather* **33**, 392–9.

PEEL, R.F. 1967: Geomorphology: trends and problems. *Advancement of Science* **24**, 205–16.

PELTIER, L.C. 1950: The geographic cycle in periglacial regions as it is related to climatic geomorphology. *Annals Association of American Geographers* **40**, 214–36.

— 1954: Geomorphology. In JAMES, P.E. and JONES, C.F. (eds.), *American Geography Inventory and Prospect* (Syracuse University Press: Association of American Geographers), 362–81.

PELTIER, L.C. 1975: The concept of climatic geomorphology. In MELHORN, W.N. and

FLEMAL. R.C., *Theories of landform Development* (Binghamton: State University of New York).

PENCK, A. and BRUCKNER, E. 1901–9: *Die Alpen im Eiszeitalter*. Leipzig: Tauchnitz, 1199 pp.

PENCK, W. 1924: *Die Morphologische Analyse*. Stuttgart: Engelhorn.

PENMAN, H.C. 1948: Natural evaporation from open water, bare soil and grass. *Proceedings Royal Society, London* (A) **193**, 120–45.

— 1950: Evaporation over the British Isles. *Quarterly Journal of the Royal Meteorological Society* **76**, 372–83.

PENNING–ROWSELL, E. 1981a: Consultancy and contract research. *Area* **13**, 9–12.

— 1981b: Fluctuating fortunes in gauging landscape value, *Progress in Human Geography* **5**, 25–41.

PENNING-ROWSELL, E.C. and CHATTERTON, J.B. 1977: *The Benefits of Flood Alleviation: a Manual of Assessment Techniques*. Farnborough: Saxon House.

PENNING-ROWSELL, E.M. and HARDY, D.I. 1973: Landscape evaluation and planning policy: A comparative survey in the Wye valley Area of Outstanding Natural Beauty. *Regional Studies* **7**, 153–60.

PERRAULT, P. 1974: *De l'origine des fontaines*. Paris. Trans. A.L.A ROQUE, New York: Hafner, 1967.

PERRY, A.H. 1981: *Environmental Hazards in the British Isles*. London: Allen & Unwin.

— 1982: Is the climate becoming more variable? *Progress in Physical Geography* **6**, 108–14.

PEWE, T.L. (ed.) 1969: *The Periglacial Environment*. Montreal: McGill-Queens University Press.

PITTY, A.F. 1982: *The Nature of Geomorphology*. London: Methuen.

PLAYFAIR, J. 1802: *Illustrations of the Huttonian Theory of the Earth*. Reprinted 1964, New York: Dover.

POPPER, K.R. 1934: *The Logic of Scientific Discovery*. London: Hutchinson.

POSER, H. 1947: Dauerfrostboden und Temperaturverhältnisse während der Würm Eiszeit im nichtvereisten Mittel und Westeuropa. *Naturwissenschaften* **34**.

PRENTICE, I.C. 1983: Postglacial climatic change: vegetation dynamics and the pollen record. *Progress in Physical Geography* **7**, 273–86.

PRICE, R.J. 1973: *Glacial and Fluvioglacial Landforms*. Edinburgh: Oliver & Boyd, 242 pp.

PRIOR, D.B. 1977: Coastal mudslide morphology and processes on Eocene clays in Denmark. *Geografisk Tidsskrift* **76**, 14–33.

PURSEGLOVE, J. and HOLMES, N. 1983: *Nature Conservation and River Engineering*. London: Nature Conservancy Council.

RAMACHANDRAN, R. and THAKUR, S.C. 1976: India and the Ganga floodplains. In WHITE, G.F. (ed.), *Natural Hazards: Local, National, Global* (New York: Oxford University Press). 36–43.

RAPP, A. 1960: Recent development of mountain slopes in Karkevagge and surroundings, northern Scandinavia. *Geografiska Annaler* **42**, 73–200.

REVITT, D.M. and ELLIS, J.B. 1980: Rain water leachates of heavy metals in road surface sediments. *Water Research* **14**, 1403–7.

RHIND, D. 1984: Geographical data-sifting. *The Times Higher Educational Supplement* No. **594**, 30.

RICHARDS, K.S. 1982: *Rivers: Form and Process in Alluvial Channels*. London: Methuen.

ROBERTS, J.M. 1980: *The Pelican History of the World*. Harmondsworth: Penguin Books, 1052 pp.

ROBINSON, G.W. 1937: *Mother Earth: being Letters on Soil addressed to Professor R.G. Stephen*. London: Murby.

RODDA, J.C., DOWNING, R.A. and LAW, F.M. 1976: *Systematic Hydrology*. London: Butterworth.

ROSTANKOWSKI, P. 1982: Transformation of nature in the Soviet Union; proposal, plans and reality. *Soviet Geography* **22**, 381–90.

RUDEFORTH, C.C. 1982: Handling soil survey data. In BRIDGES, E.M. and DAVIDSON, D.A., *Principles and Applications of Soil Geography* (London: Longman), 97–131.

RUELLAN, A. 1971: The history of soils: some problems of definition and interpretation. In YAALON, D.H. (ed.), *Palaeopedology* (Jerusalem: International Society of Soil Science and Israel Universities Press).

RUMNEY, G.R. 1970: *The Geosystem Dynamic Integration of Land, Sea and Air*. Dubuque, Iowa: Wm. C. Brown Company, 135 pp.

RUNGE, E.C.A. 1973: Soil development sequence and energy models. *Soil Science* **115**, 183–93.

RUSSELL, J.E. 1957: *The World of the Soil*. London: Collins, 285 pp.

RUSSELL, R.J. 1949: Geographical geomorphology. *Annals Association of American Geographers* **39**, 1–11.

SAARINEN, T.F. 1966: *Perception of the Drought Hazard on the Great Plains*. Chicago: University of Chicago Press.

ST ONGE, D.A. 1981: Presidential address. Theories, paradigms, mapping and geomorphology. *Canadian Geographer* **25**, 307–15.

SAVIGEAR, R.A.G. 1952: Some observations on slope development in South Wales. *Transactions Institute of British Geographers* **18**, 31–52.

—— 1965: A technique of morphological mapping. *Annals Association of American Geographers* **55**, 514–38.

SCHEIDEGGER, A.E. 1961: *Theoretical Geomorphology*. Berlin: Springer-Verlag.

SCHEIDEGGER, A.E. and LANGBEIN, W.B. 1966: Probability concepts in geomorphology. *US Geological Survey Professional Paper* **500-C**, 14 pp.

SCHNEIDER, S.H. and DICKINSON, R.E. 1974: Climate modelling. *Reviews of Geophysics and Space Physics* **12**, 447–93.

SCHUMM, S.A. 1956: Evolution of drainage systems and slopes in badlands at Perth Amboy, New Jersey. *Bulletin Geological Society of America* **67**, 597–46.

—— 1963a: The disparity between present rates of denudation and orogeny. *US Geological Survey Professional Paper* **454H**, 13 pp.

—— 1963b: A tentative classification of river channels. *US Geological Survey Circular* **477**, 10 pp.

—— 1965: Quaternary palaeohydrology. In WRIGHT, H.E. and FREY, D.G. (eds.), *The Quaternary of the United States (Princeton: Princeton University Press), 783–94.*

—— 1968: Speculations concerning palaeohydrologic controls of terrestrial sedimentation. *Bulletin Geological Society of America* **79**, 1573–88.

—— 1969: River metamorphosis. *Proceedings American Society of Civil Engineers, Journal Hydraulics Division* **95**, 255–73.

—— 1977: *The Fluvial System*. New York: John Wiley, 338 pp.

—— 1979: Geomorphic thresholds: the concept and its applications. *Transactions Institute of British Geographers* **NS4**, 485–515.

SCHUMM, S.A. and LICHTY, R.W. 1965: Time, space and causality in geomorphology. *American Journal of Science* **263**, 110–19.

SEWELL, W.R.D. (ed.) 1973: *Modifying the Weather: a Social Assessment*. Victoria, BC: Department of Geography, University of British Columbia.

SHACKLETON, N.J. and HALL, M.A. 1983: Stable isotope record of hole 504 sediments: High resolution record of the Pleistocene. *Initial Reports of the Deep Sea Drilling Project* **69**, Washington: US Government Printing Office.

SHERLOCK, R.L. 1922: *Man as a Geological Agent*. London: Witherby, 372 pp.

— 1923: The influence of man as an agent in geographical change. *Geographical Journal* **61**, 258–73.

SHERMAN, L.K. 1932: Streamflow from rainfall by the unit-graph method. *Engineering News Record* **108**, 501–5

SIMBERLOFF, D. 1972: Models in biogeography. In SCHOPF, T.J.M. (ed.), *Models in Palaeobiology* (San Francisco: Freeman, Cooper & Co.), 160–91.

SIMMONS, I.G. 1974: *The Ecology of Natural Resources*. London: Arnold.

— 1978: Physical geography in environmental science. *Geography* **63**, 314–323.

— 1979a: *Biogeography: Natural and Cultural*. London: Arnold, 400 pp.

— 1979b: Conservation of plants, animals and ecosystems. In GREGORY, K.J. and WALLING, D.E. (eds.), *Man and Environmental Processes* (Folkestone: Dawson), 241–58.

— 1980: Biogeography. In BROWN, E.H. (ed.), *Geography Yesterday and Tomorrow* (Oxford: Oxford University Press), 146–66.

SIMMONS, I.G. and TOOLEY, M.J. (eds.) 1981: *The Environment in British Prehistory*. London: Duckworth.

SIMONS, M. 1962: The morphological analysis of land forms: a new review of the work of Walther Penck. *Transactions Institute of British Geographers* **31**, 1–14.

SIMONSON, R.W. 1959: Outline of the generalized theory of soil genesis. *Proceedings of the Soil Science Society of America* **23**, 152–61.

SIMPSON, R.H. and RIEHL, H. 1981: *The Hurricane and its Impact*. Oxford: Blackwell, 398 pp.

SISSONS, J.B. 1958: Supposed ice-dammed lakes in Britain with particular reference to the Eddleston valley, southern Scotland. *Geografiska Annaler* **40**, 159–87.

— 1960: Some aspects of glacial drainage channels in Britain. Part 1. *Scottish Geographical Magazine* **79**, 131–46.

— 1961: Some aspects of glacial drainage channels in Britain Part II. *Scottish Geographical Magazine* **77**, 15–36.

— 1967: *The Evolution of Scotland's Scenery*. Edinburgh: Oliver & Boyd, 259 pp

— 1976: *Scotland*. London: Methuen.

— 1977: Former ice-dammed lakes in Glen Moriston, Ivernesshire and their significance in upland Britain. *Transactions Institute of British Geographers* **NS2**, 224–42.

SLAYMAKER, H.O. 1968: The new geography. Review of *Models in Geography. Geographical Journal* **134**, 405–7.

SLAYMAKER, H.O., DUNNE, T. and RAPP, A. 1980: Geomorphic experiments on hillslopes. *Zeitschrift für Geomorphologie* Supplementband **35**, v–vii.

SMALL, R.J. 1978, The revolution in geomorphology – a retrospect. *Geography* **63**, 265–72.

SMALLEY, I.J. and VITA-FINZI, C. 1969: The concept of 'system' in the earth sciences. *Bulletin Geological Society of America* **80**, 1591–4.

SMART, J.S. 1978: The analysis of drainage network composition *Earth Surface Processes* **3**, 129–70,

SMIL, V. 1979: Controlling the Yellow River. *Geographical Review* **69**, 251–72.

SMITH, D.E., MORRISON, J., JONES, R.L. and CULLINGFORD, R.A. 1980: Dating the main postglacial shoreline in the Montrose area, Scotland. In CULLINGFORD, R.A., DAVIDSON, D.A. and LEWIN, J. (eds.), *Timescales in Geomorphology* (Chichester: Wiley) 225–45.

SMITH, D.I. and NEWSON, M.D. 1974: The dynamics of solutional and mechanical erosion in limestone catchments on the Mendip Hills, Somerset. In GREGORY, K.J. and WALLING, D.E. (eds.), *Fluvial Processes in Instrumented Watersheds*, Institute of British Geographers Special Publication No. **6**, 155–68.

SMITH, K. 1972: *Water in Britain: a Study in Applied Hydrology and Resource Geography*. London: Macmillan.

— 1975: *Principles of Applied Climatology*. London: McGraw Hill.

SMITH, K. and TOBIN, G. 1979: *Human Adjustment to the Flood Hazard*. London: Longman.

SMITH, T.R. 1984: Artificial intelligence and its applicability to geographical problem solving. *Professional Geographer* **36**, 147–58.

SMITH, T.R. and BRETHERTON, F.P. 1972: Stability and the conservation of mass in drainage basin evolution. *Water Resources Research* **8**, 1506–29.

SNYTKO, V.A., SEMENOV, Yu, M. and DAVYDOVA, N.D. 1981: A landscape-geochemical evaluation of geosystems for purposes of rational nature management. *Soviet Geography* **22**, 569–78.

SOCHAVA, V.B., KRAUKLIS, A.A. and SNYTKO, V.A. 1975: Toward a unification of concepts and terms used in integral landscape investigations. *Soviet Geography* **16**, 616–22.

SPATE, O.H.K. 1960: Quantity and quality in geography. *Annals Association of American Geographers* **50**, 377–94.

SPEIGHT, J.G. 1969: Parametric description of landform. In STEWART, G.A. (ed.), *Land Evaluation* (Melbourne: Macmillan). 239–50.

STARKEL, L. 1976: The role of extreme (catastrophic) meteorological events in contemporary evolution of slope. In DERBYSHIRE, E. (ed.), *Geomorphology and Climate* (London: Wiley), 203–246.

— (ed.) 1981: The evolution of the Wisloka valley near Debica during the late Glacial and Holocene. Krakow: *Folia Quaternaria*, 91 pp.

— 1982: The need for parallel studies on denudation chronology and present-day processes. *Earth Surface Processes and Landforms* **7**, 301–2.

— 1983: The reflection of hydrologic changes in the fluvial environment on the temperate zone during the last 15000 years. In GREGORY, K.J. (ed.), *Background to Palaeohydrology* (Chichester, Wiley), 213–36.

STARKEL, L. and THORNES, J.B. 1981: Palaeohydrology of river basins. *British Geomorphological Group Technical Bulletin* No. **28**, 107 pp.

STATHAM, I. 1977: *Earth Surface Sediment Transport*. Oxford: Clarendon Press, 184 pp.

STEERS, J.A. 1948: *The Coastline of England and Wales*. Cambridge: Cambridge University Press.

STEPHENS, N. 1980: *Geomorphology in the Service of Man*, Inaugural lecture, University College of Swansea.

STEWART, I. 1975: The seven elementary catastrophes. *New Scientist* **68**, 447–54.

STOCKING, M.A. 1972: Relief analysis and soil erosion in Rhodesia using multivariate techniques. *Zeitschrift für Geomorphologie* **16**, 432–43.

STOCKING, M.A. 1977: Rainfall energy in erosion: some problems and applications. University of Edinburgh Department of Geography, Research Paper **13**.

— 1980: Soil loss estimation for rural development: a position for geomorphology. *Zeitschrift für Geomorphologie* Supplementband **36**, 264–73.

— 1981: Causes and prediction of the advance of gullies. In *Problems of Soil Erosion and Sedimentation*, South-east Asian regional symposium, Bangkok, 37–47.

STODDART, D.R. 1962: Catastrophic storm effects on the British Honduras reefs and cays. *Nature* **196**, 512–15.

— 1965: Geography and the ecological approach. The ecosystem as a geographic principle and method. *Geography* **50**, 242–51.

— 1966: Darwin's impact on geography. *Annals Association of American Geographers* **56**, 683–98.

— 1967a: Growth and structure of geography. *Transactions Institute of British Geographers* **41**, 1–20.

— 1967b: Organism and ecosystem as geographical models. In CHORLEY, R.J. and HAGGETT, P. (eds.), *Models in Geography* (London: Methuen), 511–48.

— 1968: Climatic geomorphology: review and assessment. *Progress in Geography* **1**, 160–222.

— 1975: 'That Victorian Science': Huxley's *Physiography* and its impact on geography. *Transactions Institute of British Geographers* **66**, 17–40.

— 1978: Progress report: biogeography. *Progress in Physical Geography* **2**, 514–28.

— (ed.) 1981: *Geography, Ideology and Social Concern*. Oxford: Basil Blackwell, 250 pp.

— 1983: Biogeography: Darwin devalued or Darwin revalued? *Progress in Physical Geography* **7**, 256–64.

STRAHLER, A.N. 1950a: Davis's concept of slope development viewed in the light of recent quantitative investigations. *Annals Association of American Geographers* **40**, 209–13.

— 1950b: Equilibrium theory of erosional slopes approached by frequency distribution analysis. *American Journal of Science* **248**, 673–96 and 800–14.

— 1951: *Physical Geography*. New York: Wiley.

— 1952: Dynamic basis of geomorphology. *Bulletin Geological Society of America* **63**, 923–37.

— 1956: The nature of induced erosion and aggradation. In THOMAS, W.L. (ed.) *Man's Role in Changing the Face of the Earth* (Chicago: University of Chicago Press), 621–38.

— 1964: Quantitative geomorphology of drainage basins and channel networks. In CHOW, V.T. (ed.), *Handbook of Applied Hydrology* (New York: McGraw Hill), 4-39 - 4-76.

— 1966: Tidal cycle of changes in an equilibrium beach, Sandy Hook, New Jersey. *Journal of Geology* **74**, 247–68.

STRAHLER A.N. and STRAHLER, A.H. 1976: *Elements of Physical Geography*. London: Wiley, 469 pp.

STREET-PERROTT, A., BERAN, M. and RATCLIFFE, R. (eds.) 1983: *Variations in the Global Water Budget*. Dordrecht: D. Reidel Publishing Company.

STUART, L.C. 1954: Animal geography. In JAMES, P.E. and JONES, C.F. (eds.) *American*

Geography Inventory and Prospect (Syracuse University Press: Association of American Geographers, 442–51.

SUGDEN, D.E. 1978: Glacial erosion by the Laurentide ice sheet. *Journal of Glaciology* **20**, 367–91.

— 1982: *Arctic and Antarctic: A Modern Geographical Synthesis*. Oxford: Blackwell, 472 pp.

SUGDEN, D.E. and JOHN, B.S. 1976: *Glaciers and Landscape. A Geomorphological Approach*. London: Arnold, 376 pp.

SUMMERFIELD, M.A. 1981: Macroscale geomorphology. *Area* **13**, 3–8.

SUMNER, G.N. 1978: *Mathematics for Physical Geographers*. London: Arnold, 236 pp.

SUNDBORG, A. 1956: The river Klaralven, a study of fluvial processes. *Geografiska Annaler* **38**, 127–316.

SUSLOV, R.P. 1961: *Physical Geography of Asiatic Russia*, translated by N.D. Gershevsky, edited by J.E. Williams, San Francisco: W.H. Freeman.

SVERDRUP, H.U., JOHNSON, M.W. and FLEMING, R.H. 1942: *The Oceans*. New York: Prentice Hall.

TANSLEY, A.G. 1935: The use and abuse of vegetational concepts and terms. *Ecology* **16**, 284–307.

TATHAM, G. 1951: Geography in the nineteenth century. in TAYLOR, G. (ed.), *Geography in the Twentieth Century* (New York: Philosophical Library), 28–69.

TAYLOR, J.A. 1984: Biogeography. *Process in Physical Geography* **8**, 94–101.

TEICHOLZ, E. and BERRY, B.J.L. 1983: *Computer Graphics and Environmental Planning*. Englewood Cliffs, NJ: Prentice Hall.

TERJUNG, W. 1976: Climatology for geographers. *Annals Association of American Geographers* **66**, 199–222.

THOM, R. 1975: Structural stability and morphogenesis: *An Outline of a General Theory of Models*, Trans. D.H. Fowler. Reading, Mass: Benjamin.

THOMAS, M.F. 1978: Denudation in the tropics and the interpretation of the tropical legacy in higher latitudes – a view of the British experience. In EMBLETON, C., BRUNSDEN, D. and JONES, D.K.C., *Geomorphology. Present Problems and Future Prospects* (Oxford, Oxford University Press), 185–202.

— 1980: Preface to HAGEDORN, H. and THOMAS, M., Perspectives in Geomorphology. *Zeitschrift für Geomorphologie* Supplementband **36**, V–VI.

THOMAS, W.L. (ed.) 1956: *Man's Role in Changing the Face of the Earth*, Chicago: University of Chicago Press.

THOMPSON, R.D., MANNION, A.M., MITCHELL, C.W., PARRY, M. and TOWNSHEND, J.R.G. 1985: *Processes in Physical Geography*. London: Longman.

THORN, C.E. (ed.) 1982: *Space and Time in Geomorphology*. London: Allen & Unwin.

THORNBURY, W.D. 1954: *Principles of Geomorphology*, New York: John Wiley & Sons.

THORNES, J.B. 1980: Structural instability and ephemeral channel behaviour. *Zeitschrift für Geomorphologie* Supplementband **36**, 233–44.

— 1983a: Geomorphology, archaeology and recursive ignorance. *Geographical Journal* **149**, 326–33.

— 1983b: Evolutionary geomorphology. *Geography* **68**, 225–35.

THORNES, J.B. and BRUNSDEN, D. 1977: *Geomorphology and Time*. London: Methuen, 208 pp.

— THORNES, J.B. and FERGUSON, R.I. 1981: Geomorphology. In WRIGLEY, N. and BENNETT, R.J. (eds.), *Quantitative Geography a British View* (London: Routledge & Kegan Paul), 284–93.

THORNES, J.E. 1981: A paradigmatic shift in atmosphere studies? *Progress in Physical Geography* **5**, 429–40.
— 1982: Atmospheric management. *Progress in Physical Geography* **6**, 561–78.
THORNTHWAITE, C.W. 1948: An approach towards a rational classification of climate. *Geographical Review* **38**, 55–94.
TIVY, J. 1971: *Biogeography. A Study of Plants in the Ecosphere*. Edinburgh: Oliver & Boyd, 394 pp.
TOOLEY, M.J. 1978: *Sea Level Changes: North-West England during the Flandrian Stage* Oxford: Clarendon Press.
TOWNSHEND. J.R.G. (ed.) 1981a: *Terrain Analysis and Remote Sensing*. London: Allen & Unwin.
— 1981b: The spatial resolving power of earth resources satellites. *Progress in Physical Geography* **5**, 32–55.
— 1981c: Prospect: A comment on the future role of remote sensing in integrated terrain analysis. In TOWNSHEND, J.R.G. (ed.), *Terrain analysis and Remote Sensing* (London: Allen & Unwin, 219–23.
TOWNSHEND, J.R.G. and HANCOCK, P.J. 1981: The role of remote sensing in mapping surficial deposits. In TOWNSHEND, J.R.G. (ed.), *Terrain Analysis and Remote Sensing* London: Allen Unwin, 204–18.
TOY, T.J. 1982: Accelerated erosion: Process, problems and prognosis. *Geology* **10**, 524–29.
TRICART, J. 1957: Application du concept de zonalité à la gémorphologie. *Tijdschrift van het Koninklijk Nederlandsch Aardrijikskundig Geomootschap*, Amsterdam, 422–34.
TRICART, J. and CAILLEUX, A. 1965: *Introduction à la géomorphologie climatique*. Paris: Sedes.
— 1972: *Introduction to Climatic Geomorphology*, translated De Jonge, C.J.K. London: Longman, 295 pp.
TRICART, J. and SHAEFFER, R. 1950: L'indice d'émoussé des galets. Moyen d'étude des systèmes d'erosion. *Revue de Géomorphologie Dynamique* **1**, 151–79.
TRIMBLE, S. 1983: A sediment budget for Coon Creek basin in the driftless area, Wisconsin 1853–1977. *American Journal of Science* **283**, 454–74.
TRUDGILL, S.T. 1977: *Soil and Vegetation Systems*. London: Oxford University Press.
— 1983: Soil geography: spatial techniques and geomorphic relationships. *Progress in Physical Geography* 7, 345–60.
TRUSOV, Y. 1969: The concept of the noosphere. *Soviet Geography* **10**, 220–36.
TUFNELL, L. 1984: *Glacier Hazards*. London: Longman.
UNVARDY, M.D.F. 1981: The riddle of dispersal: dispersal theories and how they affect vicariance biogeography. In NELSON, G. and ROSEN, D.E. (eds.) *Vicariance Biogeography: a Critique* (New York: Columbia University Press), 6–29.
UNESCO, 1977: *Hydrological maps*. Paris: Unesco, WMO.
— 1984: Climate drought and desertification. *Nature and Resources* **20 (1)**, 2–8.
UNWIN, D.J. 1977: Statistical methods in physical geography. *Progress in Physical Geography* **1**, 185–221.
— 1981: Climatology, In WRIGLEY, N. and BENNETT, R.J. (eds.), *Quantitative Geography: a British View* (London: Routledge & Kegan Paul), 261–72.
VAIL, P.R., MITCHUM, R.M., SHIPLEY, T.H. and BUFFLER, R.T. 1981: Unconformities of the North Atlantic. *Philosophical Transactions Royal Society A* **294**, 137–55.
VAN VALKENBURG, S. 1951: The German school of geography. In TAYLOR, G. (ed.),

Geography in the Twentieth Century (New York: Philosophical Library), 91–115.

VERSTAPPEN, H.Th. 1983: *Applied Geomorphology: Geomorphological Surveys for Environmental Development*. Amsterdam: Elsevier.

VINK, A.P.A. 1968: The role of physical geography in intergrated surveys in developing countries. *Tijdschrift fur Economische Soc. Geogr*. **294**, 5–68.

— 1983: *Landscape Ecology and Land Use*, translated from Dutch and edited by D.A. Davidson. London: Longman.

VISHER, S.S. 1944: *Climate of Indiana*. Bloomington: Indiana University.

VITA-FINZI, C. 1969: *The Mediterranean Valleys*. Cambridge: Cambridge University Press.

— 1973: *Recent Earth History*. London: Macmillan, 138 pp.

VON ENGELN, O.D. 1942: *Geomorphology: Systematic and Regional*. New York: Macmillan.

VREEKEN, W.J. 1973: Soil variability in small loess watersheds: clay and organic matter content. *Catena* **1**, 181–96.

— 1975: Principal kinds of chronosequences and their significance in soil history. *Journal of Soil Science* **26**, 378–94.

WALKER, D. and GUPPY, J.C. (ed.) 1978: *Biology and Quaternary Environments*. Canberra: Australian Academy of Science, 264 pp.

WALLING, D.E. 1974: Suspended sediment and solute yields from a small catchment prior to urbanisation. In GREGORY, K.J. and WALLING, D.E. (eds.), *Fluvial Processes in Instrumented Watersheds*. Institute of British Geographers, Special Publication No. **6**, 169–91.

— 1979a: The hydrological impact of building activity: a study near Exeter. In HOLLIS, G.E. (ed.), *Man's Impact on the Hydrological Cycle in the United Kingdom* (Norwich: GeoBooks), 135–52.

— 1979b: Hydrological processes. In GREGORY, K.J. and WALLING, D.E. (eds.), *Man and Environmental Processes* (London: Dawson), 57–81.

— (ed.) 1982: Recent developments in the explanation and prediction of erosion and sediment yield. *International Association of Hydrological Sciences*: Publication No. **137**.

— 1983a: The sediment delivery problem. *Journal of Hydrology* **65**, 209–37.

— 1983b: Physical hydrology, *Progress in Physical Geography* **7**, 97–112.

WALLING, D.E. and GREGORY, K.J. 1970: The measurement of the effects of building construction on drainage basin dynamics. *Journal of Hydrology* **11**, 129–44.

WALLING, D.E. and WEBB, B.W. 1983: Patterns of sediment yield. In GREGORY, K.J. (ed.), *Background to Palaeohydrology* (Chichester: Wiley), 69–100.

WALSH, R.P.D, HUDSON, R.N. and HOWELLS, K.A. 1982: Changes in the magnitude – frequency of flooding and heavy rainfalls in the Swansea valley since 1875. *Cambria* **9**, 36–60.

WALTON, K. 1968: The unity of the physical environment. *Scottish Geographical Magazine* **84**, 212–18.

WARD, R.C. 1967: *Principles of Hydrology*. London: McGraw Hill, 403 pp.

— 1971: *Small Watershed Experiments. An Appraisal of Concepts and Research Developments*. University of Hull Occasional Papers in Geography No. **18**, 254.

— 1978: *Floods. A Geographical Perspective*. London: Macmillan, 244 pp.

— 1979: The changing scope of geographical hydrology in Great Britain. *Progress in Physical Geography* **3**, 392–412.

WARD, R.G. 1960: Captain Alexander Maconochie RNKH 1787–1860. *Geographical Journal* **126**, 459–68.

WARREN, A. and GOLDSMITH, F.B. (eds.) 1974: *Conservation in Practice*. Chichester: Wiley.

WARREN, A. and GOLDSMITH, F.B. (ed.) 1983: *Conservation in Perspective*. Chichester: Wiley.

WASHBURN, A.L. 1973: *Periglacial Processes and Environments*. London: Arnold.

WATERS, R.S. 1957: Differential weathering and erosion on oldlands. *Geographical Journal* **123**, 501–9.

— 1958: Morphological mapping. *Geography* **43**, 10–17.

WATKINS, J.W.N. 1970: Against 'normal science'. In LAKATOS, I. and MUSGRAVE, A. (eds.), *Criticism and the Growth of Knowledge* (London: Cambridge University Press).

WATTS, D.R. 1966: *Man's Influence on the Vegetation of Barbados 1627–1800*. University of Hull, Ocassional Papers in Geography, No. **4**.

— 1978: The new biogeography and its niche in physical geography. *Geography* **63**, 324–37.

WEAVER, W. 1958: A quarter century in the natural sciences. Annual Report, The Rockefeller Foundation, New York, 7–122.

WEBSTER, R., LESSELLS, C.M. and HODGSON, J.M. 1976: 'DECODE' – computer program for translating soil profile descriptions into text. *Journal of Soil Science* **27**, 218–26.

WEHMILLER, J.F. 1982: A review of amino acid racemization studies in Quaternary mollusks: stratigraphic and chronologic applications in coastal and interglacial sites, Pacific and Atlantic coasts, United States, United Kingdom, Baffin Island, and tropical islands. *Quaternary Science Reviews* **1**, 83–120.

WERRITTY, A. 1972: The topology of stream networks. In CHORLEY, R.J. (ed.), *Spatial Analysis in Geomorphology* (London: Methuen), 167–96.

WETHERALD, R.T. and MATANABE, S. 1975: The effects of changing the solar constant on the climate of a general circulation model. *Journal of Atmospheric Sciences* **32**, 2044–59.

WHALLEY, W.B. (ed.) 1978: *Scanning electron microscopy in the study of sediments*. Norwich: Geo Abstracts.

WHITE, G.F. 1958: *Changes in Urban Occupance of Flood Plains in the United States*. University of Chicago, Department of Geography Research Paper No. **57**.

— 1973: Natural hazards research. In CHORLEY, R.J. (ed.), *Directions in Geography* (London: Methuen), 193–216.

— (ed.) 1974: *Natural Hazards Local, National, Global*. New York: Oxford University Press, 288 pp.

— 1974: Natural hazards research: concepts, methods, and policy implications. In WHITE, G.F. (ed.), *Natural Hazards, Local, National, Global* (New York, Oxford University Press) 3–16

WHITE, G.F., CALEF, W.C., HUDSON, J.W., MAYER, H.M., SHEAFFER, J.R. and VOLK, D.J. 1958: Changes in urban occupance of floodplains in the United States. *University of Chicago, Department of Geography Research Paper No.* **57**.

WHITE; I.D., MOTTERSHEAD, D.N. and HARRISON, S.J. 1984: *Environmental systems: An Introductory Text*. London: George Allen & Unwin.

WHITMORE, T.C, FLENLEY, J.R. and HARRIS, D.R. 1982: The tropics as the norm in biogeography. *Geographical Journal* **148**, 8–21.

WHITTAKER, R.H. 1953: A consideration of climax theory – the climax as a population and pattern. *Ecological Monographs* **23**, 41–78.

WHITTOW, J.B. 1980: *Disasters: the Anatomy of Environmental Hazards*. London: Allen Lane.

WILKINSON, H. 1963: Man and the natural environment. *Department of Geography, University of Hull Occasional Papers in Geography* No. **1**.

WILLIAMS, G.P. 1978: Hydraulic geometry of river cross sections – theory of minimum variance. *US Geological Survey Professional Paper* **1029**, 47 pp.

— 1983: Improper use of regression equations in the earth sciences. *Geology* **11**, 195–7.

WILLIAMS, G.P. and WOLMAN, M.G. 1984: Downstream effects of dams on alluvial rivers. *US Geological Survey Professional Paper* **1286**.

WILSON, A.G. 1981: *Geography and the Environment Systems Analytical Methods*. Chichester: Wiley, 297 pp.

WILSON, A.G. and KIRKBY, M.J. 1974: *Mathematics for Geographers and Planners*. Oxford: Clarendon Press, 325 pp.

WINDLEY, B.F. 1977: *The Evolving Continents*. London: Wiley.

WISCHMEIER, W.H. 1976: Use and misuse of the Universal Soil Loss Equation. *Journal Soil and Water Conservation* **31**, 5–9.

— 1977: Soil erodibility of rainfall and runoff. In TOY, T. (ed.), *Erosion Research Techniques Erodibility and Sediment Delivery* (Norwich: Geo Books), 45–56.

WISE, M.J. 1983: Three founder members of the IBG: R. Ogilvie Buchanan, Sir Dudley Stamp, S.W. Wooldridge. A personal tribute. *Transactions Institute of British Geographers* **NS8**, 41–54.

WOLMAN, M.G. 1967a: A cycle of sedimentation and erosion in urban river channels. *Geografiska Annaler* **49A**, 385–95.

— 1967b: Two problems involving river channel changes and background observations. In *Quantitative Geography Part II Physical and Cartographic Topics, Northwestern Studies in Geography* **14**, 67–107.

WOLMAN, M.G. and GERSON, R.A. 1978: Relative scales of time and effectiveness of climate in watershed geomorphology. *Earth Surface Processes* **3**, 189–208.

WOLMAN, M.G. and MILLER, J.P. 1960: Magnitude and frequency of forces in geomorphic processes. *Journal of Geology* **68**, 54–74.

WOOD, A. 1942: The development of hillside slopes. *Proceedings Geologists Association* **53**, 128–40.

WOOLDRIDGE, S.W. 1932: The cycle of erosion and the representation of relief. *Scottish Geographical Magazine* **48**, 30–6.

— 1949: On taking the ge– out of geography. *Geography* **34**, 9–18.

— 1951: The progress of geomorphology. In TAYLOR, G. (ed.), *Geography in the Twentieth Century* (London: Methuen), 165–77.

— 1956: *The Geographer as Scientist*. London: Nelson.

— 1958: The trend of geomorphology. *Transactions Institute of British Geographers* 25, 29–36.

WOOLDRIDGE, S.W. and EAST, W.G. 1951: *The Spirit and Purpose of Geography*. London: Hutchinson, 176 pp.

WOOLDRIDGE, S.W. and LINTON, D.L. 1939: *Structure, Surface and Drainage in South East England*. London: Philip.

WOOLDRIDGE, S.W. and MORGAN, R.S. 1937: *The Physical Basis of Geography*. London. Longman.

WOOLDRIDGE S.W. and MORGAN, R.S. 1951: *An Outline of Geomorphology: The Physical Basis of Geography*. London: Longman.

WORSLEY, P. 1979: Whither geomorphology. *Area* **11**, 97–101.

— 1981: Part five: Evolution. In GOUDIE, A.S., *Geomorphological Techniques* (London: George Allen & Unwin), 277–305.

WRIGHT, R.L. 1972: Some perspectives on environmental research for agricultural land-use planning in developing countries. *Geoforum* **10**, 15–33.

WRIGHT, W.B. 1937: *The Quaternary Ice Age*. London: Macmillan.

WRIGLEY, N. (ed.) 1979: *Statistical Applications in the Spatial Sciences*. London: Pion.

WRIGLEY, N. and BENNETT, R.J. (ed.) 1981: *Quantitative Geography: a British View*. London: Routledge & Kegan Paul.

YAALON, D.M. 1975: Conceptual models in pedogenesis: can soil-functions be solved? *Geoderm* **14**, 189–205.

YATSU, E. 1966: *Rock Control in Geomorphology*. Tokyo: Sozoscha, 135 pp.

— 1971: Landform materials science – rock control in geomorphology. *Proceedings First Guelph Symposium on Geomorphology*, University of Guelph, 49–56.

YE GRISHANKOV, 1973: The landscape levels of continents and geographic zonality. *Soviet Geography* **14**, 61–77.

YI FU TUAN, 1971: Man and nature. *Association of American Geographers* Commission on College Geography Resource Paper, No. **10**.

YOUNG, A. 1960: Soil movement by denudational processes on slopes. *Nature, Land* **188**, 120–2.

— 1963: Deductive models of slope evolution. *Nachrichten der Akademie der Wissenschaften in Göttingen, II Mathematisch – Physikalische Klasse* 5, 45–66.

— 1964: Discussion of slope profiles: A symposium. *Geographical Journal* **130**, 80–2.

— 1972: *Slopes*. Edinburgh: Oliver & Boyd.

— 1974: The rate of slope retreat. In BROWN, E.H. and WATERS, R.S., *Progress in Geomorphology*, Institute of British Geographers Special Publication No. **7**, 65–78.

YOUNG, A, and GOLDSMITH, P.F. 1977: Soil survey and land evaluation in developing countries. A case study in Malawi. *Geographical Journal* **143**, 407–38.

YOUNG, G.L. 1974: Human ecology as an interdisciplinary concept: a critical enquiry. *Advances in Ecological Research* **8**, 1–105.

ZAKRZEWSKA, B. 1967: Trends and methods in landform geography. *Annals Association American Geographers* **57**, 128–65.

ZEUNER, F.E. 1945: *The Pleistocene Period: its Climate, Chronology and Faunal Successions*. London: Hutchinson.

ZEUNER, F.E. 1946: *Dating the Past: An Introduction to Geochronology*. London: Methuen.

Index to authors

General index